Advance praise for *Something Th*

"The author is uniquely equipped by his formation as a scientist and his subsequent research into religious experiences to review and assess critically the confusing deluge of studies of states of the brain during such experiences. This work provides a sure-footed and clarifying guide through this particular jungle and thereby provides the necessary basis for philosophical and theological reflection on this current and very active field of study."
—Rev. Dr. Arthur Peacocke, former Warden Emeritus at the Society of Ordained Scientists and Hon. Chaplain and Honorary Canon at Christ Church Cathedral in Oxford

"Lucid and accessible, *Something There* combines the sensitivity of the past and the rigor of the scientist with the theologian's hunger for truth. Hay makes the case for placing spirituality at the very center of our quest to live meaningful and authentic lives."
—Andrew Wright, Ph.D., coordinator of the Centre for Theology, Religion and Culture (CTRC), King's College, London

"The biology of spirituality—is such a subject possible? David Hay is uniquely qualified to speak on human spirituality from a biological perspective. Having served as the director of the Religious Experience Research Unit at Oxford University, he is well acquainted with the classic studies of spirituality conducted by zoologist Alister Hardy, founder of the research unit. In this book, Hay surveys classic studies in the study of religion and spirituality, providing depth of coverage rarely seen in books of this nature. Hay also summarizes his own research work, which can be characterized by both having breadth and variety. As a complement to both the classic studies and his own work, Hay summarizes some of the best research conducted in recent years related to the spiritual experiences of humanity—including a variety of religions and, perhaps just as important, the spiritual experiences of non-religious people. The latter are important because as the influence of religion has significantly declined in much of the world, the interest in spirituality simultaneously has mushroomed. His conclusions point not only to a biological basis for spiritual experience, but also to the fact that spirituality is an important part of what makes us human, as the many first-hand accounts clearly underscore (although the possibility is admitted that some higher animals may have similar experiences). This book is poetry and science, historical and contemporary, subjective and objective, thoroughly human yet pointing beyond to the Other. A major addition to the literature on spirituality, it draws widely upon many sources—and perhaps The Source—to make a distinctive contribution to that literature."

—Donald Ratcliff, Ph.D., Price-LeBar Professor of Christian Education, Christian Formation and Ministry Department, Wheaton College, Wheaton, Illinois

SOMETHING THERE

SOMETHING THERE

THE BIOLOGY OF THE HUMAN SPIRIT

David Hay

TEMPLETON FOUNDATION PRESS
PHILADELPHIA

Templeton Foundation Press
300 Conshohocken State Road, Suite 670
West Conshohocken, PA 19428
www.templetonpress.org

2007 Templeton Foundation Press Edition
First published by Darton, Longman, and Todd, London, UK

Templeton Foundation Press helps intellectual leaders and others learn about science research on aspects of realities, invisible and intangible. Spiritual realities include unlimited love, accelerating creativity, worship, and the benefits of purpose in persons and in the cosmos.

Interior designed by Sandie Boccacci

Library of Congress Cataloging-in-Publication Data

Hay, David, 1935-
Something there : the biology of the human spirit / David Hay.
p. cm.
ISBN 978-1-59947-114-3 (pbk. : alk. paper) 1. Spirituality. 2. Religion and science. I. Title.
BL48.H367 2007
204—dc22

2006036297

07 08 09 10 11 12 10 9 8 7 6 5 4 3 2 1

Printed in the United States of America

In memory of Alister Hardy and Alan Bullock,

earnest seekers after truth

Contents

Preface

We were sitting in my office in the university and I was wondering how to put my question. Angela was a smartly dressed woman in her early forties. She had a successful career and I could tell she didn't suffer fools gladly. In the end, rather stumblingly, I said:

> What I'm doing is making a study of ... I don't know how to describe it ... it's to do with transcendent or spiritual experience, and the illustrations that I give – because it's so difficult to define – are things like the Wordsworthian idea of a presence rolling through all things, an un-named power, God or the gods. Or it could just be the wisdom of the unconscious ... the point about it is that it feels as if it is coming from somewhere else, or outside. So, my first question – is that an experience of which you are aware?

At first Angela thought the answer was 'No', but then she changed her mind. Perhaps she had had some experiences that might fit, but she had a problem with the way I expressed myself:

> I don't know whether it's [as] a woman, thinking about the experiences when I experienced them, that I was more likely to feel them as an inner force that I was drawing on, that came from inside, not from outside and whether ... that was more a male way of describing that.

Then she began to tell me about two moving incidents in her life. The first was when her son was desperately ill in hospital:

> I arrived after the diagnosis, when he was in a very delirious state, and that was the night ... that was the evening in which he would either live or die. I mean that was very clear from what everyone was saying, and so it was like touch and go. And the moment that I saw him, it was an extraordinarily unpleasant comparison to make, but when I came into the side ward where he was, he was completely naked, and he looked like a

> Belsen child … and his eyes were unseeing and he just held out his arms, and he literally couldn't see me, but he could hear me, and there was a, you know, whoomph! Mother and child … there was no way from that point on that I could leave him. And that whole night where we just slept together, completely locked together, and it just felt like, you know, that's what I mean about the force not … well, it didn't feel like outside of me. It felt like me giving my life energy to him. So whether that's the wisdom of the unconscious, or I don't know, or whatever, but it was, um, it felt to me like a conscious willing, of summing up every ounce of energy I had to will him into life, to pull him back from death. And I knew that I would succeed.

I was a little unsure of what I was hearing so I asked Angela, 'Why do you think this experience fits with what I am interested in?'

> Because I think the energy source that I was tapping into for that time was something that I don't normally tap into … I want to call it energy, and it feels like that to me, as though there was some sort of energy source that I was pulling out of – and I'm gesturing towards the floor as if it was – you know, from the earth itself.

The mood of our conversation had been darkly intense, but then Angela went on to talk about a very different experience that happened when she was on holiday with her family in Portugal:

> We were on a deserted beach early one morning in the Algarve and, as you do in a family, you ramble your own separate ways along the rocks, and then come together, and then ramble off again. And we were all round this rock pool and in one moment (because it was quite a small rock pool), streaked – and I heard this, it was like 'Shooomf!' like that – this octopus, and the moment that he went 'Shooomf' it was just exquisitely beautiful, the movement of the thing, it crossed the rock pool and then proceeded to flop out of the side of it in a vain bid to reach the sea. And we were all completely … we'd seen it all at the same moment, you know. And the concern was how it would not make it to get to the sea, other than by divine intervention, i.e. us rescuing it, but it was quite a sort of hefty thing, and we'd only got a bucket that big, but the kids went and got some sea

> water and I managed to get it on a spade and then put it in the water, and then deliver it to the sea. At the moment it went in, it went 'Shooomf!' again. It just felt like a wonderful ecstatic moment … of perceiving the beauty of nature.

Did Angela's answers correspond with what I was looking for? She seemed to be unsure about whether the force was coming from inside or outside. She made no mention of God except for a joking reference to her family as 'divine intervention'. And to be practical, what bearing does such apparently fragile and evanescent experience have on the concrete world of economic and political reality that presses in upon us every day of our lives?

Over the past decade there has been an astonishing surge of interest in spirituality throughout much of the Western world. Opinions differ on its significance. I am a zoologist by profession and my thoughts have evolved out of a lengthy personal and professional association with my fellow zoologist the late Alister Hardy, at one time Regius Professor of Natural History at my *alma mater*, Aberdeen University and latterly Professor of Zoology in Oxford. In 1969, shortly after he retired, Hardy founded the Religious Experience Research Unit (RERU) at Manchester College in Oxford, and it was there that I began working with him in 1974. Like Hardy, I am a committed Darwinian, but also like him, I am a religious believer. To live on that boundary at this point in our cultural history is an uncomfortable, sometimes painful experience, but the rawness and uncertainty I have felt whilst staying there is the source of whatever scientific and religious creativity I possess.

As someone who has spent his entire career in empirical science, I have benefited greatly from the freeing up of the mind associated with the European Enlightenment. The Enlightenment also has a down side, including a propensity to ignore or positively reject the reality of our spiritual nature, bringing in its wake a widespread sense of ultimate meaninglessness. The most obvious victims are the religious institutions that are currently in a state of severe crisis in large parts of the Western world, but I believe that the political damage in terms of loss of social cohesion or 'social capital' is equally great. It is therefore of interest that in recent years a considerable body of evidence has been accumulating in both the physical and social sciences suggesting that our spiritual nature is real and not illusory. Or as many of the people I have spoken with during my research put it, 'there is something there'.

In this book I have attempted to place the evidence in the context of our Western cultural history and to think through the general implications for human well-being.

The immediate trigger was the surprising findings of a two-year investigation I directed at Nottingham University and completed in the millennium year. My colleague Kate Hunt and I were having conversations about their spiritual lives with a group of local Nottingham people, selected on the criterion that they had consciously chosen not to belong to any formal religious organisation. Their comments strengthened my sense, built up during thirty years' research, that spiritual awareness is a necessary part of our biology, whatever our religious beliefs or lack of them.

Over the years I have been grateful for the support of very many people, too numerous to mention. Nevertheless I wish to record in particular my debt to Alister Hardy, who first invited me to come to work with him in Oxford and in doing so reoriented my academic career. I am also grateful to the late Alan Bullock, who for a time chaired the advisory council of RERU and continued to take an active interest in its work almost to his dying day. The unstinting support of both men gave me the courage to carry on working in a controversial field during some personally trying periods when I was very relieved to have their friendship.

With regard to the research with non-churchgoers that triggered the writing of this book it is a pleasure to record my thanks to those organisations and individuals who decided that my idea was worth supporting financially: the Bible Society, the Church of England, the Methodist Church, the British Province of the Jesuits, George Carey (at the time Archbishop of Canterbury), James McGuinness, Bishop of Nottingham, Professor Andrew M. Greeley of the University of Chicago, and a charitable trust that wishes to remain anonymous. The project has also benefited from its advisers who, whilst they may not always have shared my personal perspective, willingly offered their expertise: Professor Robert Dingwall, Professor Margaret Donaldson and Gordon Heald. In his role as Director of the Opinion Research Business (ORB) Gordon was also in charge of a national survey commissioned by the BBC in connection with a series of programmes entitled *Soul of Britain*. Through his agency we were able to include a set of questions about spirituality in the survey and I am grateful to both ORB and the BBC for their help. ORB was also responsible for recruiting and organising the focus groups in Nottingham that were the subject of the qualitative side of our research. Kate Hunt, my research assistant throughout the period of the project, carried out a major part of the fieldwork and I am very grateful to her for her commitment and the creativity she brought to our work together. Particularly in chapter 3 I have borrowed freely from the case study material she wrote up for our research report in 2000. We both owe a great debt of gratitude to the Nottingham citizens who permitted us to spend so much of their time in conversation about the life of the spirit.

As a zoologist keeping track of all that has been going on in this field I have frequently needed to depend on the expertise of others. They include: Colwyn Trevarthen and Nikos Evangelou for advice on neurophysiology, Peter Hay for help with nuclear medicine, Grace Davie for her special knowledge of the religious situation in Europe, Larry Wilde for help with Marx's views on human essence, Michael Barnes SJ for tutoring me on the philosophy of Emmanuel Levinas, John Milbank for a lunchtime chat on boundary issues in science and theology, and Philip Endean SJ for his expertise on Karl Rahner. For the past ten years I have been invited to give an annual lecture series in the Institute for the Study of Religion at the Jagiellonian University in Krakow and have benefited greatly from trying out my ideas with some very able students as well as having enlightening discussions on the psychology of religion with Halina Grzymala-Moszczynska and Pawel Socha. Similarly, on regular visits to the University of Warsaw I have learned from the enthusiasm and originality of Tomasz Ochinowski.

A special thank you is due to Adrian Bullock for sterling advice on publishing matters. I also thank Ron Garner, Gerard W. Hughes SJ, Arthur and Rosemary Peacocke, and Denis Rice for reading and commenting on draft chapters. Jane Hay's love and intuitive skill repeatedly helped me to recover my composure at those points where I came to a halt or became despondent about the book. Finally, I wish to express my delight at returning to my *alma mater* as an honorary member of the Department of Divinity and Religious Studies. My colleagues have gone out of their way to welcome me, a zoologist with only an amateur knowledge of theology. In particular I want to mention John Drane who invited me into the department in the first place (although he has since left it), and John Swinton, with whom I continue to share the ideas in this book.

DAVID HAY
Centre for Spirituality and Health
King's College, Aberdeen University

Part 1

CONTEXT

Chapter 1

THE MOUNTAINS OF THE MIND

Man that is born of woman has but a short time to live, and is full of misery.

1662 Prayer Book

O the mind, mind has mountains; cliffs of fall
Frightful, sheer, no-man-fathomed.

Gerard Manley Hopkins[1]

ESTRAGON (GIVING UP): *Nothing to be done.*

Samuel Beckett, *Waiting for Godot*[2]

THE SPIRIT OF AN ERA

Certain people come to represent the spirit of their time. Maybe the Beatles or Elvis did that for the pop world of the 1960s and 1970s. On a cold evening in London in the 1950s the Irish playwright Samuel Beckett had already gone a step further. He put in a nutshell the emotional culmination of four centuries of European culture. The setting of this *tour de force* was the Arts Theatre Club in Great Newport Street on 3 August 1955. After troubles with the Lord Chamberlain over censorship, Beckett's new play was about to open privately with Peter Hall directing. The weather had been drab all day, so it was not a bad time to get out of the cold and into the warmth of a theatre. Anyone present on that basis and hoping for a comfortable evening was to be disappointed. As the curtain rose, this is what confronted them.

On a bare stage stand two shabbily dressed men. Bowler-hatted clowns, they might be mistaken for Laurel and Hardy.[3] They are Vladimir and Estragon, the central characters in *Waiting for Godot*.

Although there is a path running through their Wasteland they do not venture along it, but stay rooted to the spot. It seems they are immobilised by hope, the hope that someone called Godot will arrive and ... what? Give a meaning or an explanation of their situation? Tell them what to do? Get them out of their hopelessness? Who knows? At any rate here-and-now they are in a state of absolute dependency on a mysterious figure who may or may not exist.[4] Those members of the audience who have the stamina to remain in their seats until the final curtain find out that nothing changes and Godot still hasn't come. They are left with the strong suspicion that he will never come.

On that opening night, and for many succeeding nights, most of the audience were unimpressed. Many walked out, while others stayed to jeer – they couldn't make head or tail of it. I remember being puzzled myself the first time I saw the play. Yet Beckett repeatedly asserted that the theme of *Waiting for Godot* is simple and straightforward. His claim to transparency is somewhat disingenuous, for his words confront us with human depths where few would care to remain for long. The psychological world occupied by the central characters is one that has regressed back to a childlike directness of feeling, as hinted at by the use of their pet names, Didi and Gogo. Their straightforwardness has a primordial quality, as if signalling the announcement of an embodied, biologically rooted knowingness existing before all theology, all philosophy, all scientific investigation, or any kind of extended thought whatsoever. The common response of those first audiences was denial, in the psychiatric sense of that word. People either avoided Beckett's version of the truth by refusing to understand at the conscious level, or they ran away.

In spite of the temptation to shut out Beckett's message as unrecognisable, his language is indeed perfectly familiar. Inevitably, but paradoxically, the naked passions of the characters in *Waiting for Godot* are clothed in the forms of European culture, particularly its Christian beliefs[5] – inevitably, because Beckett was a European; paradoxically, because how can language ever do more than hint at the primordial?

Beckett was unusually well acquainted with Christian culture. His mother, who was a devout member of the Church of Ireland, brought him up in a suburb of Dublin.[6] Their relationship was troubled and intense, and for the rest of his life he continued to immerse himself in the classical literature of Christianity, particularly the works of Dante. An interviewer once asked him if a Christian interpretation of *Godot* was justified, to which he replied, 'Yes, Christianity is a myth with which I am perfectly familiar. So naturally I use it.'[7] Beckett's allusions to Christianity are not always reverent, but they are always serious and self-referential. His preoccupations are identical to those of the devout Christian believer – meaning, hope and despair, suffering and the shortness of life.[8] During one knock-about exchange Estragon makes a reference to Jesus Christ. Vladimir exclaims 'Christ! What has Christ to do with it? You're not going to compare yourself to Christ!' Estragon replies: 'All my life I've compared myself to him.'

Many critics have dwelt upon this pervasive use of Christian imagery in Beckett's writing, some of them claiming that his message is religious. It has even been suggested that he was awarded the 1969 Nobel Prize for Literature on the mistaken assumption that his writing was a defence of religion. But Beckett was absolutely without any form of religious belief. The concluding words of his novel *The Unnameable*, express his views succinctly: 'Where I am, I don't know, I'll never know, in the silence you don't know, you must go on, I can't go on, I'll go on.'[9]

One way of interpreting *Godot* is to see it as a snare to catch out unwary people and summarily demolish their illusion that they do know where they are.[10] Although Beckett's stock of ideas came from the European tradition, he insisted that there is no solid reason to suppose that this or *any* inherited point of view corresponds with reality. His personal axioms included the belief that we are forever alone, that language ultimately fails to communicate, that broad generalisations (metanarratives) about the nature of reality are unwarranted; that they are even a kind of violence done to the unique world of the individual. Hence all that a writer or artist who feels driven to express themselves can do is to speak as concretely and simply as possible about their own experience of life.

THE MARGINALISING OF RELIGION

Although Beckett expressed his pain in a novel way, what he said is not new. Loss of coherent meaning lies at the core of an austere tradition traceable back to at least the sixteenth century in Europe.

From time to time the distress of it is caught in the writings of those who deal in naked feeling – that is, the poets. There is a parallel emotion running through from John Donne's early seventeenth-century lament,

> And freely men confess that this world's spent,
> When in the planets and the firmament
> They seek so many new: they see that this
> Is crumbled out again t'his atomies.
> 'Tis all in pieces, all coherence gone;
> All just supply, and all relation:[11]

to Matthew Arnold's verse written in 1867:

> Ah, love, let us be true
> To one another! for the world, which seems
> To lie before us like a land of dreams,
> So various, so beautiful, so new,
> Hath really neither joy, nor love, nor light,
> Nor certitude, nor peace, nor help for pain;
> And we are here as on a darkling plain
> Swept with confused alarms of struggle and flight,
> Where ignorant armies clash by night.[12]

W. B. Yeats continues with similar vehemence in 1921,

> The falcon cannot hear the falconer;
> Things fall apart; the centre cannot hold;
> Mere anarchy is loosed upon the world ... [13]

In all three poems the loss of coherence is expressed as a spiritual or religious loss and it is a commonplace that this is associated with ideas that came to fruition during the European Enlightenment.[14] Just how certain aspects of the Enlightenment made religious belief difficult for Europeans is an issue I will explore in detail later (see chapter 9). At this point it is sufficient to be reminded that the legacy of religious scepticism is inherited to some degree by everyone in the continent of Europe, not to speak of those other parts of the world that have been strongly influenced by European ideas. In every

Westerner who is at all socially aware there resides either the conviction, or the knowledge of other people's conviction, that religion is a dubious affair. Even those who have withdrawn into a religious ghetto to protect their beliefs are conscious of the critics who have triggered their wish for radical disengagement from the dominant culture.

Beckett is a supremely powerful spokesman on behalf of that underlying scepticism. In the 1950s, at what now seems like a pivotal moment, with the skill of a psychoanalyst he brought to consciousness the sense of incoherence that had been growing for centuries, but in the main was ignored or repressed in people's everyday lives. He even got it headlined in the popular media. It is as if, with the honesty and naïvety of a small boy, he blurted out an unpalatable intuition; there *is* no ultimate meaning. Large numbers of our contemporaries, when they have the strength to explore that suppressed world, share the opinion that the meanings that traditionally clothed our existence are simply not there.

The loss of religious meaning is publicly displayed in the statistics of decline in the churches. Britain is fortunate in having a relatively good set of figures summarising the changing involvement of people in religious institutions over the past 150 years. Though interpretation of the data is difficult and controversial, the waning of these institutions is unmistakable. It has been examined in detail by students of secularisation theory (which predicts that as society becomes more rationally ordered, the influence of religion will decline).[15] The picture that emerges is one of increasing numbers of people drifting away from what they see as no longer credible, hence irrelevant to their lives. During the second half of the twentieth century the collapse appears to have accelerated.[16]

The story is well known so I shall confine myself to offering only a few reminders of the argument. Steve Bruce, the leading British proponent of secularisation theory, reflecting on several different analyses of the 1851 Census of Religious Worship, concludes that probably between 40 per cent and 60 per cent of the adult population of Britain were regular church attenders in that year.[17] By 1998, according to figures gathered by Peter Brierley,[18] the proportion had dropped to 7.5 per cent. Between 1900 and 2000 the aggregated figures for church membership fell from an optimum of 27 per cent of the national population to 10 per cent.[19] The story is much the same for religious professionals. In 1900 there were over 45,000 clergy in the United Kingdom. By the year 2000, at a time when the national population had almost doubled, the number of clergy had fallen to just over 36,000. Professor Bruce points out that to maintain equivalence with 1900, there should be something like 80,000 clergy today.[20] Even more startling, Brierley calculates that attendance of children at Sunday School dropped from a maximum of half the population in 1920 to a miniscule 4 per cent in 2000.[21] In England in 1900, according to Brierley, 67 per cent of all weddings took place in an Anglican church, whilst by 2000 the figure had fallen to 20 per cent.[22] Finally and most devastatingly, during the last decade of the twentieth century people were detaching themselves from the mainstream churches at a spectacular rate. The statistics recorded in *Religious Trends 1999/2000*[23] show that regular church attendance in Britain fell from 4.74 million in 1989 to 3.71 million in 1998, a drop of more than 20 per cent in ten years. Extrapolating those figures forward, one would have to conclude that there will be virtually no Christian institutional presence in Britain by the year 2050. In other words, if we take religious adherence as our measure of a belief in ultimate meaning, then there is overwhelming evidence that Beckett's message has hit home.

RELIGIOUS AND SPIRITUAL EXPERIENCE

Although this book is about spirituality, I will inevitably have a great deal to say about religion. The two subjects are very closely related, to the extent that for many people they are synonymous. What then, in the light of the statistics just quoted, are we to make of the dramatic changes in report of religious or spiritual *experience* in Britain during the latter part of the twentieth century? In 1987, along with Gordon Heald (who at the time was director of Gallup Poll in Britain), I published the results of a survey of reports of such experience in the UK.[24] The figures showed that 48 per cent of the national sample felt they were personally aware of this kind of experience in their lives. In the year 2000, in association with the BBC's *Soul of Britain* review of the spiritual state of the nation, I had the opportunity to have another look at the question. Again I used the skills of Gordon Heald, who by this time was running his own polling organisation, the Opinion Research Business (ORB). I wondered what had happened over the years since 1987, approximately the same period of time during which church attendance had dropped by 20 per cent. I was curious to see whether there was a parallel fall in positive response to questions about religious and spiritual experience. As far as possible I decided to repeat the 1987 enquiries in the new survey, though omitting two of the original questions.[25]

I was astonished when I received the results, to the extent of telephoning the ORB office to make sure there had not been a mistake. Over those 13 years there had been an almost 60 per cent increase in the positive response rate. The figures suggest that around three-quarters of the national population are now likely to admit to having had one of these experiences. The great majority of these people are of course not regular churchgoers. And ever since the millennium, if the mushrooming of references on spirituality on the Internet is anything to go by (according to Google at the time of writing, it had risen to over 36,000,000)* there is no sign of the interest dying down.

But what do people mean when they say they have had a spiritual or religious experience? I am able to give a substantial answer because of the work of the Oxford zoologist, Alister Hardy, of whom I shall have more to say in later chapters. In 1969 Hardy founded the Religious Experience Research Unit (RERU)[26] with the purpose of making a scientific study of the nature, function and frequency of reports of religious experience in the human species. He saw his initial task as rather like that of the Victorian naturalists, gathering examples of different specimens and classifying them, in preparation for the creation of a theory of religious experience. In response to advertisements he placed in the media, Hardy accumulated an archive of several thousand personal descriptions sent in to the Unit by members of the general public. These narratives were replies to variants of the following question:

> Have you ever been aware of or influenced by a presence or a power, whether you call it God or not, that is different from your everyday self?

Hardy was soon to find that the classification of the responses was more difficult than he expected, primarily because there is a fundamental difference between sorting physical organisms like animals

* Currently (as of July 2006) the figure is 104,000,000.

and plants, and applying the same method to written accounts of experience. In the latter case a multitude of personal, psychological, social and political influences affect the way people put their experience into words, making classification a more puzzling and uncertain undertaking.

In spite of the complications, several attempts have been made to organise the material in Hardy's archive.[27] The simplest, and therefore the crudest method, is the one Gordon Heald and I used to prepare the questionnaire for our 1987 survey. We wanted to ask about different subcategories of experience. To be useable in a large-scale national poll the questions have to be clear-cut, straightforward and as far as possible, unambiguous. We decided that it was not practicable to offer a list of more than a few alternatives, so we carefully reviewed the archive held in the RERU office and identified the eight commonest types of experience recorded. These were inserted in our poll. For the sake of clarity I will present the same classification here, along with a warning that this simplification is of course an *over*simplification.

In my illustrations I have intentionally chosen vivid descriptions because they are helpful to convey the strong feeling that typically lies behind them. But there is a snag about using extracts like this because most people's accounts are simple and down-to-earth. Hardy wanted to stress this ordinariness, in contrast to the two previous best-known studies of religious experience prior to his work, which both emphasised extraordinary states of consciousness. William James, first professor of Psychology at Harvard University, is the acknowledged founding father of the modern psychological study of religious experience. He achieved this position through his Gifford Lectures, delivered in Edinburgh University in 1901–2 and published as *The Varieties of Religious Experience.*[28] The lectures intentionally laid stress on extreme examples of experience because of James' belief that psychological phenomena are most easily recognisable in their acute form. Similarly, in 1917, the German philosopher and theologian Rudolf Otto wrote a highly influential book on the experience of transcendence, *Das Heilige.*[29] It is a dramatic work in which he focuses upon the awe-inspiring forms that such experience can take, drawing many of his examples from the Bible.

Table 1: *Frequency of Report of Religious or Spiritual Experience in Britain for the Years 1987 and 2000*

	1987	**2000**
A patterning of events	29%	55%
Awareness of the presence of God	27%	38%
Awareness of prayer being answered	25%	37%
Awareness of a sacred presence in nature	16%	29%
Awareness of the presence of the dead	18%	25%
Awareness of an evil presence	12%	25%
Cumulative Total	**(48%)***	**76%**

* This includes totals for respondents to two additional questions asked in 1987 about 'a presence not called God' (22%) and 'awareness that all things are One' (5%), i.e. the total of 76% for the year 2000 is quite likely to be relatively speaking an underestimate.

Bearing my proviso in mind, here are the classified extracts from Hardy's archive.

AWARENESS OF A PATTERNING OF EVENTS/ SYNCHRONICITY

The commonest kind of experience reported in Britain is the recognition of a transcendent providence: a patterning of events in a person's life that convinces them that in some strange way those events were meant to happen. In the millennium year survey I mentioned above, 55 per cent of the national sample recognised this in their own lives. This is a 90 per cent rise compared to the response when the question was asked in 1987 (see Table 1). Sometimes these events have the startling characteristics of what the psychologist C. G. Jung called 'synchronicity', that is, a 'meaningful coincidence' or a cluster of events that do not appear to have any causal connection with each other, yet have a meaningful relationship.[30]

My first example will illustrate what I mean. A young woman is giving an account of her religious search:

> ... while walking home one dark night I reflected how my search was going and, rather sadly, felt that, like Thomas, I must have proof and without that I would have to say that I did not

> believe in God. Deep in thought, I looked up at the night sky, which was filled with hundreds of stars. Wildly, I threw the silent call upwards, 'Prove it!' Hardly had the words been formed than a bright star sped across the sky. Before it died away, another star had begun to traverse the darkness. And there, just for a moment, an enormous cross blazed in the heavens like a personal signature. I was filled with awe and a certain terror at the power that I saw unleashed …

Here we see at once the importance of the cultural context in which an experience (any kind of experience) occurs for the construction of meaning, since this woman is aware of the symbolic significance of the cross in Christianity. Nevertheless, whilst a sceptic would dismiss the coincidence as meaningless, she is convinced that her experience is not merely coincidental, since she adds,

> My husband died last year at the age of 39 years with cancer. While I nursed him a friend said to me, 'I don't know how you can believe in God.' The question surprised me, for once you know there is a God the question of belief ceases to exist.

The next example has a rather less obviously culturally constructed content, but the symbolic importance of coincidence is experienced as even more melodramatic. The incident occurred at a time when the informant's life was in pieces and she had decided to kill herself:

> … at that moment I let out a loud challenge into that dark and lonesome night, into that desolation of land and soul and I shouted: … IF THERE IS SUCH A THING AS A GOD THEN SHOW YOURSELF TO ME – NOW … and at that very instant there was a loud crack, like a rifle shot [coming from the bedroom] … I stumbled through the open door to my bedroom. I fell into the bed shaking and then something forced my eyes upward to the wall above my bedside table and where I had a very small photograph of my father hanging … The picture had gone – I just looked at the empty space … but in looking closer I saw the photograph, face down on the little table and the narrow silver frame was split apart, the glass broken and from behind the cardboard on the back there had slipped out … the last letter [my father] had written me … When I picked up that letter and read over and over the words of this beloved caring

> father of mine, I knew that was HIS help to me, and God answered me directly in the hour of this soul being in anguish.

These two examples are described in spectacular terms and experienced as such. Much more commonly, people speak of coming to recognise an unfolding pattern in their lives that has not been dictated by their personal choice, as for example in the selection of a career. Almost without exception this configuration is interpreted as something 'given', though not necessarily with an overtly religious connotation. The next example, however, is from a committed religious believer, and here we can see a move towards St Ignatius Loyola's dictum of 'seeing God in all things':

> The experiences of the last six months have … confirmed my deep conviction that God is directly and indirectly guiding my life … as well as being absolutely convinced of Divine guidance in the larger issues of my life, I feel the guidance strongly even in some of the smaller events … the pattern of my life seems to me to be a mosaic, in which everything, including seeming disasters, eventually turns to good …

AWARENESS OF THE PRESENCE OF GOD

Many people feel they have been aware of the presence of God. We know from our research that this can often be when they are very happy. At the other end of the scale, people talk of being aware of God when they are deeply distressed. In the latter case typically nothing changes in their physical circumstances, but the experience of God's presence places them in a larger context of meaning which helps them to bear their suffering. In the year 2000 poll, 38 per cent of the sample said they had personal awareness of such a divine presence – a 41 per cent rise on 13 years previously. Here is an example of someone experiencing feelings of joy in the presence of God, yet anxious about her sanity:

> I was looking after the Friends Meeting House high on a spur of the forest, and sleeping on a camp bed in the sitting-room of the dwelling next door. One night I awoke slowly at about one o'clock to a feeling of absolute safety and happiness; everything in the world around me seemed to be singing 'All is very well'. After an almost unbelieving (sic) few minutes I got up and went

> to the window and saw the valley filled with the love of God, flowing and spreading from the roadside and the few houses of the village. It was as though a great source of light and love and goodness was there along the valley, absolutely true and unchangeable. I went outside and looked down over the hedge, and the light and assurance were most truly there; I looked and looked, and, to be honest, I was not thankful, as I should have been, but trying to absorb the awareness of safety and joy so deeply that I would never forget it.

The writer goes on to say that the following day, in someone else's house, she picked up a magazine that lay 'open at an article on just such religious experiences as [I] had had the previous night'. Synchronicity again. She interpreted it as a reassurance that she was 'quite sane'. As we shall see, anxiety about insanity is a common accompaniment of these experiences.

Some of the most interesting examples in Hardy's archive are memories from childhood.[31] The freedom for experience of this type in children is often attributed to naïvety and a misunderstanding of causality. In a later chapter I will question this dismissal and offer an alternative account based on recent empirical research. Here is an example of someone recollecting a spontaneous experience of being in God's presence when she was a young girl:

> My father used to take all the family for a walk on Sunday evenings. On one such walk, we wandered across a narrow path through a field of high, ripe corn. I lagged behind, and found myself alone. Suddenly, heaven blazed upon me. I was enveloped in golden light, I was conscious of a presence, so kind, so loving, so bright, so consoling, so commanding, existing apart from me but so close. I heard no sound. But words fell into my mind quite clearly – 'Everything is all right. Everybody will be all right.'

The writer connects her comment with the best-known saying of the fourteenth-century English mystic, Julian of Norwich, 'All shall be well, and all shall be well, and all manner of thing shall be well.'

Surprise is very characteristic, as in the next quotation, which is of interest because the writer includes a reflection on the use of metaphor to mediate his experience:

> The experience itself is very difficult to describe. It took me completely by surprise. I was about to start shaving at the time, of all things. I felt that my soul was literally physically shifted – for quite a number of seconds, perhaps 15 to 20 – from the dark into the light. I saw my life, suddenly, as forming a pattern and felt that I had, suddenly, become acquainted with myself again after a long absence – that I was, whether I liked it or not, treading a kind of spiritual path, and this fact demanded me to quit academics and enter social work ... I must stress here that prior to this experience I used never to use the words such as 'soul' or 'salvation' or any such 'religiously coloured' words. But in order to make even the slightest sense of what happened to me I find it imperative to use them. Looking back it does seem as if I saw a kind of light, but I think that this might have been a metaphor I coined immediately after the experience.

The next illustration is an example of the many accounts of the presence of God that are associated with times of severe distress:

> I had an experience seven years ago that changed my whole life. I had lost my husband six months before and my courage at the same time. I felt life would be useless if fear were allowed to govern me. One evening with no preparation, as sudden and dynamic as the Revelation to Saul of Tarsus, I knew I was in the presence of God, and that he would never leave me nor forsake me and that he loved me with a love beyond imagination – no matter what I did.

AWARENESS OF A PRESENCE NOT NAMED

Rather more than a fifth of our 1987 sample referred to this kind of experience. We were not able to include a question on it in our millennium poll.[32] That is unfortunate, because it is an increasingly important category in an age of religious decline. Luckily there are numerous examples in Hardy's archive. People are often lost for words, and when they do find words they are not sure that they convey the experience. Evidently the term 'God' is not always appropriate for the writers, partly because they feel uneasy with religious language. The inarticulacy associated with any experience of a transcendent presence, and the uncertainty when words *are* found, is beautifully brought out in the next account. It is also of a childhood memory, from a man in his fifties:

> As a child (not younger than 6, not older than 8) I had an experience which nowadays I consider as kindred, if not identical with those experiences related by Wordsworth in *The Prelude*, Bk I, lines 379–400.[33] The circumstances were: dusk, summertime, and I one of a crowd of grown-ups and children assembled round the shore of an artificial lake, waiting for full darkness before a firework display was to begin. A breeze stirred the leaves of a group of poplars to my right; stirred, they gave a fluttering sound. There, then, I knew or felt or experienced – what? Incommunicable now, but then much more so. The sensations were of awe and wonder, and a sense of astounding beauty … that child of 6 or 7 or 8 knew nothing of Wordsworth or about mysticism or about religion.

My former colleague Edward Robinson[34] had the opportunity to talk further with this man about what his experience meant to him. He told Robinson,

> It's very difficult to say that it revealed – what? The existence of infinity? The fact of divinity? I wouldn't have had the language at my command to formulate such things, so that if I speak about it now it is with the language and ideas of a mature person. But from my present age, looking back some half a century, I would say now that I did then experience – what? a truth, a fact, the existence of the divine. What happened was telling me something. But what was it telling? The fact of divinity, that it was good? Not so much in the moral sense, but that it was beautiful, yes, sacred.[35]

Did this man mean 'God'? I don't know, and it is not clear that the person himself feels able to give an unequivocal answer to that question either. A similar uncertainty appears in the next extract, because the informant has consciously withdrawn from the religious institution in which he was nurtured:

> Since about the age of 6 I have had an awareness of a higher power. At all times I am aware of this power, which is as real to

> me as any in the physical world. In this sense I live in two spheres of influence. When I am tranquil, as in bed late at night, I place my problems before this higher power and I am shown the way to solve them … Originally a Catholic, since 12 I have belonged to no organised religion whatsoever. I belong to no group.

Sometimes there is no equivocation, as in the next brief extract, which sounds self-contradictory, perhaps further underlining the difficulty with the use of religious terminology in a secular age:

> I … know that since I concluded some years ago that my mind could not accept a personal God … I seem to have become more aware of this all pervading power which to me is strength, comfort, joy, goodness …

AWARENESS OF PRAYER BEING ANSWERED

In great unhappiness or fear, many people, including those who are uncertain about God's existence, turn to prayer for help. A total of 37 per cent of those questioned in the millennium survey felt they had received such help – a 40 per cent increase on 1987. My first example is a description of a modern vision. It is of great interest because of the common perception that such experience is symptomatic of mental disturbance, though this was not the opinion of the informant. His understanding of the shaping role of culture in his experience is clear, for he remarks 'I realise that the form of the vision and the words I heard were the result of my education and cultural background', but this does not lead him to dismiss it. At the time of the incident he had been a psychiatric patient for three years during which he underwent numerous electric shock treatments for schizophrenia. He interprets his experience not as a symptom of his illness, but rather as the trigger for his recovery from psychosis. He writes:

> At one time I reached utter despair and wept and prayed God for mercy instinctively and without faith in reply. That night I stood with other patients in the grounds waiting to be let into our ward. It was a very cold night with many stars. Suddenly someone stood beside me in a dusty brown robe and a voice said, 'Mad or sane, you are one of my sheep.' I never spoke to

anyone of this, but ever since, 20 years, it has been the pivot of my life.

My next example is an illustration of prayer as something involuntary, emerging from severe distress and followed by religious awareness. The informant was watching through the night with his dying father:

> I was stretched out hardly a foot away from my father as his life slipped from his body, and came to the shocking conclusion that I was of very little help to him. He lay there not making much sound, just enough to make me aware that he was in distress. Every so often he pulled at the covers, trying to get out of bed. I was miserably conscious of each movement as he struggled towards the edge of the mattress and was in danger of crashing onto the floor. Again and again I got up, walked round to the other side of the bed and tried to settle him back as gently as I could. I thought to myself, this is the worst night of my life. I found I was praying. Not words. Just a despairing reaching towards God to help me through the night. Then, slowly, an extraordinary change began to take place. I became more and more strongly aware of God's presence filling the room, indescribably powerful and (does this make any sense to you?) drawing my father and me, all things, together in a vast, rich harmony. Then it seemed as if something like a hard crust was dissolving or falling away inside me. I knew what was happening. The wounded relationship was being tenderly uncovered and healed. I was filled with joy. The rest of what I experienced is beyond words …

In the following example, a prayer for enlightenment is answered, and once again the cultural context is striking. I do not know if the writer was aware of the lines in George Herbert's hymn,

> A servant with this clause makes drudgery divine.
> Who sweeps a room as for thy laws makes that and the
> action fine.

> I prayed with unashamed sincerity that if God existed, could He show me some sort of light in the jungle. One day, I was

> sweeping the stairs, down in the house in which I was working, when suddenly I was overcome, overwhelmed, saturated ... with a sense of most sublime and living love. It not only affected me, but seemed to bring everything around me to life. The brush in my hand, my dustpan, the stairs, seemed to come alive with love. I seemed no longer me, with my petty troubles and trials, but part of this infinite power of love, so utterly and overwhelmingly wonderful that one knew at once what the saints had grasped. It could only have been a minute or two, yet for that brief particle of time it seemed eternity ...

A SACRED PRESENCE IN NATURE

Another commonly reported experience is an awareness of a sacred presence in nature, akin to Wordsworth's description of a presence that 'rolls through all things' in his lines written above Tintern Abbey. A total of 29 per cent of the millennium sample felt that they had had this kind of experience – an 81 per cent rise since 1987. In the next example, what begins as an awareness of a presence in nature and the hearing of God's words to Moses out of the burning bush, ends as the experience of mystical union with God:

> I was staying in Ireland in a cottage by the sea, with a beach, sand dunes and mountains. Walking through the dunes to call the family home from their fishing in the river, I came to the hollow I had walked through many times. This time I was halted by a voice saying clearly 'Take off your shoes, the ground on which you stand is holy ground.' I had no shoes on, but I was compelled on to my knees and then into a crouch so that I was as close as I could be to the ground. Then a tremendous silence came around me; I almost felt it touched me, I was enclosed in it. Yet I could hear the insects, bees, beetles, ants, etc. in the small flowers in the short grass, and I was one with them, creatures and flowers. I could hear the sheep and the breakers beyond on the beach, and I was one with them and the sea. Rivers, and for some reason the Victoria Falls (which I have never seen), came into my mind, and I was one with all waterfalls, all trees, all living things everywhere. A farmer's wife in the valley had just had a baby and I was one with them, and the old woman on the mountain who was dying and her relatives who

> were with her before they left for America, and I was one with them. Not only 'one with', somehow I was them. Then I thought, 'God is here, with me and in me, the Creator', and for that moment I was one with and in God.

The next excerpt seemingly belongs to the same universe of discourse, but at the time it was without the overt religious content of the previous example:

> I did not attribute any great significance to these experiences: they were an expression of my ecstatic love for what Wordsworth calls 'natural objects', not utterly different from the ecstasy of sexual love. I did not think of them in terms of union with God, for instance, until much later. I used to be puzzled by the way this experience would come unheralded, and in the most unlikely places – not, for instance, in rose plot, fringed pool, fern'd grot – but in a bus or by a dustbin; but I did not think a lot about it or try to give a meaning to it until I read Wordsworth, and, later still, various books on mysticism.

AWARENESS OF THE PRESENCE OF THE DEAD

A surprisingly large number of people, 25 per cent of the national sample in the year 2000, felt they had been in touch with someone who has died – this is a 38 per cent rise since 1987. This kind of experience is commonly reported in association with immediate grief following the death of a loved companion or relative, as in the following incident:

> After the sudden death of my husband about nine years ago, I had several experiences, which proved to me that there is life after death. I am not a spiritualist nor a churchgoer, but I try to follow Jesus, and I am a great believer in meditation as a way to God. After his passing, I both saw and spoke to my husband and held his hand. This hand was strong and not at all ghost-like, nor was his appearance. I was alone at the time, so no medium there to act as a link. Probably this is not a detail to prove God's existence, but to me, it indeed did.

Sometimes the dead person offers comfort because of some other grief or anxiety, as in the next example:

> One day I got a phone message to say that my elder son, who was at [a private school] had been taken to hospital that morning with polio. As I lay across the kitchen table in complete anguish and despair, I distinctly felt my grandmother's hand laid on my shoulder and I had a feeling of complete serenity. My grandmother had died when I was 11. She was a very religious woman and I adored her. I just could not worry after this happening, and my son managed to throw off the germ before the paralysis set in.

AWARENESS OF AN EVIL PRESENCE

A quarter of all the people interviewed in 2000 felt they had been aware of an evil presence – a rise of over 100 per cent since 1987. These experiences are qualitatively quite different from all the other categories we have looked at, in that they are associated with a sense of great dread and unhappiness, as described here by a man in his mid-twenties:

> A year and a half ago I was asleep in the night and woke very suddenly and felt quite alert. I felt surrounded and threatened by the most terrifying and powerful presence of evil. It seemed almost physical and in a curious way it 'crackled', though not audibly ... I felt it was a manifestation directed very personally at me by a power of darkness.

Quite often, even when people say they are not religious, they turn to traditional religious symbols in an attempt to allay their fear. During a lengthy car journey through Europe, a woman and her companions had arranged to stay the night in a large town:

> We had all been travelling many hours and were hot, sticky and tired. On entering the room I felt a most terrible chill, a fear I had never known. I am afraid I cannot put into words what exactly I felt, only to say that some terrible presence was in this room also ... On the bedside table beside me was a Bible: although I am of the Jewish faith and not religious, this, even so, made me feel at that time very close to God ... I do not believe one has to be religious to speak with God. But the Bible made me feel strong, made me feel that whatever was in this room I could fight and that God would fight alongside me.

Sometimes an encounter with a very physical manifestation of evil can, so to speak, trigger off a vocation to combat it. Here a GP recalls an incident when she was a girl:

> Something came to me when I was thirteen years old. I saw a picture of a pile of Jews' bodies waiting to be bulldozed into a mass grave in Germany during the Second World War. Of course I was shocked, but the 'whatever it is' took the opportunity to pound my brain from the inside for twelve hours. I can remember thinking 'I don't know what you want. What am I to do?' At the end of the twelve hours I thought – though I can't remember why – that maybe I should choose a career as a doctor.

AWARENESS THAT ALL THINGS ARE ONE

The notion of coherence, unity, shading into the mystical 'All is one' appears in a relatively small number of reports, and in our 1987 survey only 5 per cent responded positively to this question. It was this consideration that led me to omit it from the millennium survey. These accounts are nevertheless extremely interesting, since a number of professional students of mysticism have suggested that the experience of unity is the most fundamental form of mystical experience.[36] My first example is very like one I quoted from in the category of 'a presence in nature' except that the person does not claim to be aware of the presence of God:

> One afternoon I was lying down resting after a long walk on the Plain ... The grass was hot and I was on an eye level with insects moving about. Everything was warm, busy and occupied with living. I was relaxed but extraneous to the scene. Then it happened: a sensation of bliss. No loss of consciousness, but increased consciousness ... I could feel the earth under me right down to the centre of the earth, and I belonged to it and it belonged to me. I also felt that the insects were my brothers and sisters, and all that was alive was related to me, because we were all living matter that died to make way for the next generation ... And I felt and experienced everything that existed, even sounds and colours and tastes, all at once, and it was all blissful ... I had a conviction that a most important truth

> had been enunciated: that we are all related – animal, vegetable and mineral – so no one is alone. I have never forgotten this experience.

Finally, an aesthetically rich description which turns into a very pure experience of unity:

> I was walking across a field, turning my head to admire the Western sky and looking at a line of pine trees appearing as black velvet against a pink backdrop, turning to duck egg blue/green overhead, as the sun set. Then it happened. It was as if a switch marked 'ego' was suddenly switched off. Consciousness expanded to include, be, the previously observed. 'I' was the sunset and there was no 'I' experiencing 'it'. At the same time – eternity was 'born'. There was no past, no future, just an eternal now … then I returned completely to normal consciousness finding myself walking across the field, in time, with a memory.

INSTITUTION AND EXPERIENCE PART COMPANY

The remarkable rise in reports of spiritual or religious experience in Britain during the last decades of the twentieth century is extraordinary and takes some explaining. My own guess is that in reality there has been no great change in the frequency with which people encounter a spiritual dimension in their lives. What is probably altering is people's sense that they have social permission for such experience. Somehow or other (perhaps through the influence of postmodernism, which I shall discuss later), there is a growing feeling that it is acceptable to admit to spiritual awareness, though it is still something most people feel quite deeply embarrassed about. The freeing up may be associated with a breakdown of the traditional assumption that the formal religious institution and spirituality, whilst perhaps not synonymous, are inseparable.

No doubt most people still connect the two, but there is evidence of an accelerating sense of disjunction between institution and personal experience right across the Western world. It is risky to make general assertions on the basis of statistical findings in one country. Nevertheless it seems that Britain is not unique in this respect. My colleague David Tacey at La Trobe University in

Melbourne has detected an almost identical pattern of decreasing church attendance and rising report of spiritual experience in Australia. He comments strikingly that:

> In Australia, Catholic students who abandon formal worship within eighteen months of graduating from school amount to a staggering 97 per cent of the student body. These are not figures that any institution would be proud of, and consequently they are not broadcast.[37]

At the same time his report on his own small-scale studies with his students makes the point about spirituality:

> In March 1998, I surveyed 50 of my students who had enrolled in one of my literature and psychology courses. An impressive 47 students indicated that personal spirituality was a major concern in their lives, while only two students said that religion was important. In 2002, I surveyed 125 students in my undergraduate subject, and 115 expressed personal concern for 'spirituality', while only about ten said they were pleased to be designated as following one of the religions.[38]

There are hints of a somewhat similar phenomenon in the European countries examined at intervals by the European Study of Values (ESV). Some years ago the English sociologist of religion, Grace Davie, developed the concept of 'believing without belonging',[39] and this seems to have relevance to the data gathered by the ESV. The French sociologist Yves Lambert, commenting on these surveys, notes that for the nine countries considered as a whole, the rate of self-definition as 'a religious person' among young people with no formal religion went up from 14 per cent in 1981 to 22 per cent in 1999; belief in God from 20 per cent to 29 per cent and belief in life after death from 19 per cent to 28 per cent. He adds,

> The development of this autonomous, diffused religiosity, detached from Christianity, which appeared in the survey of 1990, is the most unique phenomenon. This 'off-piste' religiosity is illustrated mainly through variables that are less typically Christian: 'taking a moment of prayer, meditation, contemplation or something like that'; belief in 'a life after death' (which can include diverse conceptions such as belief in

> reincarnation); belief in God as 'some sort of spirit or life force'; and being led to 'explore different religious traditions' rather than 'stick to a particular faith'.[40]

Even in the United States, which is often cited as a Western country that is bucking the trend towards secularisation, there is evidence accumulating that a split has appeared between religion and spirituality. In a pioneering piece of research published in 1997, Brian Zinnbauer and his colleagues[41] showed that even though the Christian churches in the US are numerically relatively five or six times as strong as in the UK, there is a parallel increasing tendency to make a distinction between 'spirituality' and 'religion'.

I believe that this shift in the relationship between religion and spirituality has been implicitly there for a very long time in Western consciousness, like a ghostly presence nudging people out of alienation, the feeling of being far from home in a cold universe. I think I see the shift in Karl Marx. As a young man in the 1840s Marx interpreted the institutional spirituality of his time as itself alienating, a false resolution of the injustices created by class society. But he also spoke of 'spirit' in highly positive terms. To criticise a situation as 'spiritless', as he does when he refers to religion as 'the spirit of a spiritless situation' is an attack on religion, but surely not on whatever he means by 'spirit'. There lies the problem. So far I have dodged the question of definition because, as is already evident, the answer is not straightforward.

Chapter 2

'UNFUZZYING THE FUZZY': WORKING TOWARDS A BIOLOGY OF THE SPIRIT[1]

The spirit of the Lord has been given to me, for he has anointed me, he has sent me to bring good news to the poor, to proclaim liberty to captives and to the blind new sight, to set the downtrodden free, to proclaim the Lord's year of favour.

Isaiah 61:1–2 (quoted by Jesus in the synagogue at Nazareth)[2]

Years ago I recognized my kinship with all living beings, and I made up my mind that I was not one bit better than the meanest on earth. I said then, and I say now, that while there is a lower class, I am in it, and while there is a criminal element I am of it, and while there is a soul in prison, I am not free.

Eugene Debs, American Socialist[3]

A DYSFUNCTIONAL FAMILY OF MEANINGS

On my shelves I have two kinds of books on spirituality. One sort is easy to pick out. Scattered here and there amongst brightly-coloured paperbacks are hardbacks with dark and battered covers, usually published in the early years of the twentieth century or before. I acquired most of them from the libraries of seminaries or convents that were closing down. They are a poignant illustration of the statistics of institutional decline.

All of those old books assume that terms like 'spirit' and 'spirituality' refer simply and obviously to Christianity. They are manuals of instruction for the Christian who wishes to advance in the devout life. Quite often they contain touching reminders of their original owners in the form of prayer cards and holy pictures, or what used to be called 'spiritual bouquets'. A card left between the pages of my copy of B. W. Maturin's *Laws of the Spiritual Life*[4] gives hints that it once belonged to Sister Mary Josephine, a Poor Clare nun living in a convent near Manchester in England at the end of World War I.

Whilst Maturin's book urges virtue upon the reader, his idea of the spiritual life is remarkably privatised, even individualistic. Each of the main chapters is a meditation on one of the *Beatitudes* preached by Jesus in his Sermon on the Mount. Chapter 5 is on the fourth beatitude, 'Blessed are they that hunger and thirst after justice, for they shall have their fill.' Maturin interprets this primarily as a plea for fasting and abstinence:

> Blessed are they whose spiritual appetites have not been destroyed by overfeeding on earthly or heavenly things, or whose spiritual efforts have not been so unwisely or unhealthily made as to interfere with its hunger and thirst after God.[5]

At least from the perspective of faith this could be construed as wise advice but, remarkably, what is entirely lacking throughout the chapter is any obvious reference to social justice. The absence becomes glaring when we compare Maturin's words with Jesus' announcement of his mission in the synagogue at Nazareth. There, the spirit of the Lord proclaims a communitarian form of justice: release for the poor and rejected from the bonds of poverty, imprisonment, blindness and persecution.

Nevertheless, political turmoil cannot have been very far from Sister Mary Josephine's consciousness. Also preserved between the pages of *The Laws of the Spiritual Life* is a postcard dated 6 January 1919. The last sentence from her correspondent reads, 'What times we live in', referring to the devastations of World War I.[6] At the same date on the other side of the Atlantic the socialist leader Eugene Debs was being held in jail for sedition because of his opposition to the war. Debs is remembered to this day as a man with a Christ-like temperament, known on occasion for returning home from winter meetings minus his overcoat because he came across a poor person without one, or giving the last of his food to someone who was hungry.[7] He cast in his lot with the despised and dispossessed without any overt reference to the Spirit, yet his actions are surely more akin to Jesus' proclamation in the synagogue than Maturin's reflections on the fourth beatitude. I do not know how Sister Josephine responded to the call for us to hunger and thirst after justice, nor do I mean to criticise her in making this juxtaposition. It does however highlight the fuzziness of meaning surrounding the term 'spirituality'.

Surely the unquestioning assumption that 'spirituality' refers only to religion can't be right? Even in contexts where that *is* the major assumption, there are exceptions. For example, in an influential published series on world spirituality,[8] amongst the many volumes dealing with specific religions there is a 562-page book entitled *Spirituality and the Secular Quest.*[9] Its chapters include discussions of the spiritual aspects of the New Age, holistic health practices, psychotherapy, feminism, gays and lesbians, scientific enquiry and ecological activism. None of these are necessarily linked to formal religion, though of course they could be. And what is the meaning of 'spirituality' in relation to a slender, exhausted-looking man I remember being pointed out to me at a cocktail party with the remark, 'He's very spiritual, you know'? Apparently the basis for this identification was that he enjoyed poetry and string quartets.

Again, what are we to make of a section headed 'Spirit' that until recently appeared regularly in the weekend magazine of my daily newspaper? According to the copy of the *Guardian Weekend* in our house at the time I was writing these paragraphs, Spirit includes:

> ***Fashion and beauty*** Why preppy boys will be sitting pretty this summer. Plus: Jess Cartner-Morley; Shopping life; the new black … ***Wellbeing*** The dos and don'ts of toilet etiquette … ***Relationships*** All the men I've never slept with; Almost a grown up; We love each other … ***Pets*** Cat stuck up a tree again? Think twice before dialling 999 … [10]

In the same issue of the *Guardian* magazine there is an article about political mutations in the New Left of the British Labour Party; also a piece on the former BBC journalist Rageh Omaar's experience of the nightmare of the Iraq war. These items certainly have relevance to those held in prison and the sufferings of the poor and downtrodden, though they are not listed under 'Spirit'.

There are still other connotations, some of them radically at odds with religion. Karl Marx, from whom Eugene Debs took some of his inspiration, famously saw religion as false consciousness, 'the spirit of a spiritless situation'.[11] Accordingly Debs' fellow American and committed Marxist, Nancy Bancroft, insists in a chapter in the book *Marxism and Spirituality* that whilst spirituality is important to her, she has no truck with religious belief. She writes bitterly of the theories of those Communists who hob-nob with religion as 'castrators of Marx'.[12] Nevertheless she offers a Marxist version of spirituality, constructing it on the basis of Marx's term 'species being' which refers to:

> ... the deepest center or spirit of humankind as a collective. The term asserts that there is no division between individual and society: 'Human' means precisely social ... we complete our individual and species character only by social interaction over time ... Species being in its full sense cannot obtain until we have eliminated class and ended every kind of social division.[13]

In spite of Bancroft's contempt for Christian-Marxist alliances, her rhetoric carries obvious echoes of Jesus' declaration in Nazareth.

THE AMBIGUITY OF LANGUAGE

Uncertainty about what counts as spirituality is further compounded by ambiguities of language. Even if we limit ourselves to studying experience that fairly obviously falls within the category of traditional religion, there are so many ways that people express their spirituality that we might wonder if they are truly distinct or merely variations due to differences in life history or culture. The issue is to do with the function of the metaphors we use when we describe an experience. Does our language in a sense construct the experience we are describing or is there some more fundamental preverbal awareness underlying the description?

The problem becomes particularly acute in those instances where we are unsure whether or not someone is talking about or hinting at the presence of God. One of the letters sent in to the Religious Experience Research Unit in Oxford will help to make the point a little clearer: This correspondent is recalling a painful incident:

> The following occurred at a time when I had no feeling for religion. It was not the result of religious ecstasy or a joyous

> heightening of the spirit. A certain event had hurt and humiliated me. I rushed to my room in a state of despair, feeling as worthless as an empty shell. From this point of utter emptiness it was as though I were caught up in another dimension. My separate self ceased to exist and for a fraction of time I seemed part of a timeless immensity of power and joy and light. Something beyond this domain of life and death. My subjective and painful feelings vanished. The intensity of the vision faded, but it has remained as a vivid memory ever since. *Years later I read of Pascal's moment of illumination and was amazed at the similarity of mine.* [Italics added]

The final sentence in this quotation is a reference to the seventeenth-century French mathematician Blaise Pascal's 'second conversion', an experience that overwhelmed him on the night of 23 November 1654. It was so important to Pascal that he wrote a description on a piece of parchment and sewed it into his clothing, where it was found after his death. It reads,

> FIRE,
> God of Abraham, God of Isaac, God of Jacob, not of the philosophers and savants.
> Certitude. Certitude. Feeling. Joy. Peace.
> God of Jesus Christ.
> My God and thy God
> 'Thy God shall be my God'
> Forgetfulness of the world and of everything except God.
> He is to be found only in the ways taught in the Gospel.
> Grandeur of the human soul.
> Righteous Father, the world hath not known Thee, but I have known Thee.
> Joy, joy, joy, tears of joy.
> I have fallen from Him.
> 'They have forsaken Me, the fountain of living waters'.
> 'My God wilt Thou forsake me?'
> May I not fall from Him for ever.
> This is life eternal, that they might know Thee, the only true God,
> and Jesus Christ whom Thou has sent.

Jesus Christ
Jesus Christ
I have fallen away: I have fled from Him, denied Him,
crucified Him.
May I not fall from Him for ever.
We hold Him only by the ways taught in the Gospel.
Renunciation, total and sweet.
Total submission to Jesus Christ and to my director.
Eternally in joy for a day's exercise on earth.
I will not forget thy word. Amen.[14]

When we compare these two texts, separated by more than three hundred years, it is quite obvious that at one level the modern description is not in the slightest bit like Pascal's account. Pascal's parchment is passionately Catholic and floridly devotional, mentioning the words 'God' or 'Jesus' sixteen times. The modern account is spare, minimal and restrained and the writer positively denies that his experience has any reference to God or religious belief. So why does this correspondent say he was 'amazed at the similarity' between his experience and that of Pascal?

One published commentary on both these passages asserts explicitly that 'it is plain that Pascal is not trying to record the happening as it occurred, he is using it as a peg on which to hang an affirmation of Catholic loyalty.'[15] The commentary seems to be asking, 'What really happened, as opposed to what Pascal misleadingly *tells* us happened?' with the implication that the modern account is a good deal nearer the truth. The man who sent in the modern narrative to the Religious Experience Research Unit doesn't seem to be bothered by that question, and assumes that he and Pascal are really talking about the same experience with different words. This implies that he sees himself (and Pascal) as having to use language, with all its cultural limitations, to communicate something of a 'primordial' human experience, a bodily awareness or 'felt sense' that is there before words or thinking happens. Furthermore, that awareness was already pregnant with meaning before he opened his mouth or sat down to write a letter about it to the Research Unit.

Another way of interpreting these two descriptions is to dismiss them as the meaningless result of hallucination or self-delusion. Samuel Beckett would probably have supposed the latter, whilst feeling compassion for the loneliness of spirit that engendered the fantasy.[16] More radically still, is such experience symptomatic of mental disturbance, or, let us be frank, a form of insanity? Pascal's vision has been explained by Vítězslav Gardavsky, a Czech Marxist writer, as being nothing but the result of the emotional shock he experienced when he accidentally fell under the hooves of a shying horse.[17]

THE CULTURAL SPLIT

Underlying the fuzziness and dispute about the meaning of the term 'spirituality' is a longstanding split in Western culture. At this moment the division enters every vein of our creative experience, that is, the way we go about explaining the mysterious reality in which we find ourselves. The split can be put very simply. Whatever may be the case now, historically Western society has been constructed and pivoted on the life and teachings of its culture hero, Jesus Christ, who claimed to speak with the utmost familiarity with God and urged his followers to do the same. Direct encounter with the sacred, what Mircea Eliade calls 'hierophany'[18] is the traditional motivator of Europe, reinforced in a multitude of hymns, as in the fourth-century Liturgy of St James:

> Let all mortal flesh keep silence,
> And with fear and trembling stand;
> Ponder nothing earthly-minded,
> For with blessing in his hand,
> Christ our God to earth descendeth,
> Our full homage to demand.[19]

Or, thirteen hundred years later, in the words of the poet George Herbert:

> Teach me my God and King
> In all things thee to see
> And what I do in anything
> To do it as for thee.[20]

Not only is European history littered with speech acts directed towards God, multitudes of people, obscure as well as famous, have claimed to have encountered God at the heart of their lives. The champions of the culture, the saints after whom streets, churches, hospitals, schools and entire cities are named, from St Petersburg to Peterborough, are traditionally people to whom God has spoken particularly clearly. In concert with this, devout Christians are urged in the Scriptures to listen to what God has to say to them in their daily lives: they are advised to place themselves in the presence of God; to wait upon God; to see God in all things, or as the Jesuit poet Gerard Manley Hopkins put it, to see the world 'charged with the grandeur of God'. To summarise, in traditional Western culture the most practical of all human experiences, because it is an encounter with the source of all being, is the encounter with God.

On the other hand, when we consider the thought arising from the Enlightenment, we find that it culminates in a central assertion from the philosopher Immanuel Kant. He claimed that we can have no direct experience of *noumena*, that is, things in themselves as they really are, as opposed to the given world of physical appearances, or what we can perceive with our senses. It follows that for Kant it is perfectly all right for people to think about God (and Kant thought a great deal about God) but there is no way we can encounter God directly. In other words, neither for ordinary people nor for philosophers has God any reality beyond being the subject of a theoretical belief. If people do make any further claim to personal experience of the supposed divine presence, they are deluding themselves. It is only a short step from this conclusion to Ludwig Feuerbach's famous projection theory of God. In his book *The Essence of Christianity*, published in 1841, Feuerbach claimed that the Christian religion (and by implication, every theistic religion) is a projection onto an imaginary God of all the best qualities of human beings, leaving them helpless and degraded, or as he put it: 'the more empty life is, the fuller, the more concrete is God'.[21]

TRANSCENDING THE ARGUMENTS: A BIOLOGICAL BASIS FOR SPIRITUALITY

So there we have the split at its sharpest. Ranged against each other are the ancient view of encounter with God as the most directly practical of all experiences because it is a meeting with the 'really real'; and the Enlightenment view of God as the most remotely theoretical of all intellectual fantasies. Over the last few centuries the strength of this latter claim has put religion out of court for increasing numbers of people in the Western world. Because of its close association with religion, spirituality too became suspect.

One person who did not feel intimidated by the critique of religious experience (hence spirituality) was the zoologist Alister Hardy. His evolutionary account of the origins of religious experience is pivotal to a new way of thinking about spirituality that will help us to get beyond the religious/secular or believer/sceptic split that I have been discussing. His insights are closely tied in with his biography, which has within it the familiar tensions produced by the division.

Hardy was born in Nottingham in England in 1896 and sent for his secondary education to Oundle School in Northamptonshire. Like most English boarding schools, Oundle also included a great deal of sport in the curriculum, a fact that troubled Alister's mother. She got her GP to write to the headmaster instructing him that her son was of a delicate constitution and must not be allowed to play games. Instead, whilst his classmates were engaged on the interminable sports afternoons, Alister was to be sent on walks in the surrounding countryside. It was during these prolonged solitary wanderings that the adolescent Hardy discovered he was a nature mystic. He was very diffident about it and was not able to admit publicly to the intensity of his feeling until very late in life, in an unpublished autobiographical piece written when he was in his eighties:

> Just occasionally, when I was sure no-one could see me, I became so overcome with the glory of the natural scene that for a moment or two I fell on my knees in prayer – not prayer asking for anything, but thanking God, who felt very real to me, for the glories of his kingdom and for allowing me to feel them. It was always by the running waterside that I did this, perhaps in front of a great foam of Meadow Sweet or a mass of Purple Loosestrife.

Hardy entered Exeter College in Oxford University in 1914, where he read zoology. His tutor was Julian Huxley, grandson of the formidable T. H. Huxley, known colloquially as Darwin's Bulldog for his fierce defence of the theory of Natural Selection. Hardy and Julian Huxley were eventually to become lifelong friends, but like many a religiously inclined student of biology before and since, Alister found his first contact with Darwinism upsetting. He quickly realised that there was a distressing gulf between his understanding of the religious doctrines that were the context of his spiritual life and the evolutionary theory he was being taught by Huxley.

In some ways the problem was familiar enough. Darwin had upset (and continues to upset) conservative religious believers because his ideas contradict a literal reading of the account of creation in the Book of Genesis. If, as Darwin suggested, animals and plants had evolved through natural selection into a multitude of species over many millions of years, this could not be reconciled with the story of the seven days of creation in the Bible. The conservatives were also disturbed because Darwin's explanation of the adaptations of animals and plants to their environment undercut a well-known argument for the existence of God. This so-called 'argument from design' infers from the fact that reality is ordered and abides by physical laws that there must be an orderer and law-giver, that is, God.

Whilst he was a theology student at Christ's College Cambridge, Darwin had read the required university text on the Design argument, William Paley's *Natural Theology*.[22] Paley's famous opening analogy is to compare the pattern apparent in nature with the design of a watch. As the existence of a watch implies a watchmaker, so the adaptations of living things to their environment require the existence of a divine designer. The book impressed Darwin, so it is ironic that he himself brought about the downfall of Paley's elegant line of reasoning with the publication of *The Origin of Species* in 1858. His proposal of natural selection as the driving force producing adaptations of species to their environment dispensed with the necessity of a 'divine watchmaker', or, as Richard Dawkins bluntly put it in 1986, the watchmaker is blind.[23]

But Alister Hardy's difficulty was nothing to do with worries about the literal truth of the Bible, nor was it to do with philosophical arguments for the existence of God. He was puzzled by the importance religious people seemed to give to the argument from design. Why did they make such a fuss about an abstract speculation when the sacred was staring them in the face, so to speak? Nor was Alister upset in the slightest by evolutionary theory. On the contrary, he was entranced by its elegance and explanatory power. But its juxtaposition with the criticisms of evolution by conventional

Christian believers (and he saw himself as that at the time) made him feel very uneasy. Worse still, secular critics of religious belief shared some of the views of religious conservatives. Both constituencies seemed to be convinced that if Darwin's demolition of the argument from design was correct, then the existence of a God was no longer credible. Hardy felt that they were arguing over an irrelevancy. The question that concentrated his mind was, 'How could the God of whom he had first become so intensely aware as a boy in the Northamptonshire countryside be dismissed on the basis of an academic argument?'

THE GIFFORD LECTURES

It is a question that never left him, and after retiring from a career that had led him to the heart of the scientific establishment as Professor of Zoology at Oxford, he gave all his energy to it. When he was approaching the age of 70 he was invited to give the Gifford Lectures at the University of Aberdeen.[24] These annual lectures were first set up towards the end of the nineteenth century in each of the four ancient Scottish Universities (Aberdeen, Edinburgh, Glasgow and St Andrews) by a bequest from a senior Scottish lawyer, Adam, Lord Gifford. Gifford's main requirement was that the lectures should be concerned with what we have been discussing, that is, the demonstration of the existence and attributes of God from an examination of design in nature.

That was not quite what Hardy had in mind for his lectures. He took the opportunity to put into words what he had been brooding upon for the whole of his professional life. It amounted to this. In spite of the great emphasis that philosophers and theologians have given to the argument from design, there is something perverse about coming to an abstract conclusion about the direct experience of a transcendent presence that people were already aware of anyway. For them to set aside their own direct experience of that 'something' in favour of a philosophical conclusion, would be rather like a man who is bothered by an uncertainty as to whether the friend sitting opposite him is really there. Having decided on the basis of a logical argument that his friend is indeed slouched in the armchair, he can relax and safely proceed to invite him for a game of snooker or a drink. That kind of behaviour might be fine for bored philosophy undergraduates with nothing better to do on a rainy day, but even philosophers in their everyday lives act on the assumption that their direct intuitions are, in most cases, reliable.[25]

Hardy expressed his conviction that all of us as members of the species *Homo sapiens* have the potential for spiritual awareness. Amongst the thousands of metaphors human beings have used to describe it we might say: a presence rolling through all things, an unnamed power, God or the gods, a power coming from the unconscious, or (as Angela put it in the Preface) energy drawn from the earth itself. Hardy argued that this awareness is like a sense; it is there because it has an important function. It has indeed been 'naturally selected' in the process of evolution because it helps us to survive.

The first series of lectures was delivered in Marischal College, Aberdeen, in the session of 1963–64 and published as *The Living Stream: A Restatement of Evolution Theory and its Relation to the Spirit of Man.*[26] The major thrust was to question whether the process of natural selection is as mechanistic as many biologists claimed. In other words, is the watchmaker really as blind as some suppose? As a standard illustration of this supposedly blind process, school children are often told a 'just-so' story of how the giraffe got its long neck. Giraffes feed off the leaves of tall trees, hence when leaves are in short supply those giraffes with longer necks will be able to reach the leaves higher up the tree. In very difficult feeding conditions these animals are more likely to endure the famine and thus survive and pass on the genes for long necks to their offspring. On this thesis, the process of natural selection is directed entirely by mechanical factors outside the control of the animal. But to concentrate on external environmental causes as the main agents of selection was, in the opinion of Hardy, to ignore a much more significant evolutionary dynamic in most animals, especially in the higher mammals, including *Homo sapiens.* He was referring to the 'internal' influence of conscious choice on the part of the animals themselves, or as he described it, behavioural selection.

One of his illustrations came from the evidence being gathered during the 1950s of a new habit appearing amongst certain birds – the opening of milk bottles, first the cardboard tops, then the metal tops – spreading apparently by copying, right through the tit populations of Europe. Given the permanence of this change of habit, in due course any members of the tit population with a gene complex giving a beak slightly better adapted to such activity would have a better chance of survival than those less well equipped. When he presented this idea at a meeting of the Linnean Society in London, some wit reflected on what might happen if the metal tops were made thicker, in order to combat the birds. Would they develop beaks shaped like tin openers? Exactly right, said Hardy. Active choice (and in the case of the human species, conscious choice) is, he claimed, in a majority of cases the directing agent and precursor of natural selection. At the time that he gave his lectures Hardy had to make this claim carefully. To the inattentive ear it sounded rather like a version of a discredited explanation of evolution proposed by the French biologist Jean-Baptiste Lamarck shortly after 1800, thus predating Darwin. Lamarck is popularly known for suggesting that bodily adaptations to the environment will be passed on to offspring. It is the cliché that the blacksmith's muscles, developed by slaving for years at the anvil, will be passed on genetically to his children. They will not. Since those days the interaction of social and biological evolution has been dramatically developed, most significantly in William Durham's magisterial book *Co-Evolution,* first published in 1991.[27]

This point is of considerable importance in combating the view that evolution is an entirely mechanical process. It led into Hardy's other major argument, expounded in his second series of lectures, published as *The Divine Flame.*[28] Hardy agreed with Edmund Burke that we are 'religious animals'. In his view, as part of the process of consciously investigating their environment, forerunners of the human species discovered an awareness of a transcendent presence that met them in a way that set the experience apart from the every-day. One might guess it had parallels to Hardy's own boyhood experience in the fields of Northamptonshire.

How far back in evolutionary terms this consciousness might stretch is not clear, but Hardy certainly assumed that it was not confined to the human species. There have been sporadic scientific speculations about the existence of spiritual awareness in other animals[29] since shortly after the publication of *The Origin of Species.*[30] The psychologist Wolfgang Köhler's field reports in the 1920s of communal dancing he had observed in apes have been interpreted as possible religious ritual.[31] Other observers have given graphic descriptions of apparently religious behaviour in baboons in response to the rising and setting of the sun.[32] In his book *The Biology of God*, published in 1975,c[33] Hardy himself speculated about the religious quality of the devotional behaviour of dogs.

These conjectures may be thought far-fetched, but there is rather stronger evidence for the existence of spiritual experience in *Homo sapiens neanderthalensis.* One well-known example is a Neanderthal grave discovered in the Shanidar cave in Northern Iraq by Ralph Solecki in 1960. The grave, which Solecki judged to be 60,000 years old, was found to contain large amounts of pollen from bright flowers,[34] implying a reverence for the deceased and some sense of life beyond death. In spite of subsequent critiques,[35] remarkable finds like this make it at least plausible that spiritual awareness existed in pre-modern humans.

On the other hand, the vividness of his own experience meant that Hardy found Feuerbach's notion of religion as a projection simply too thin and speculative to be plausible. Nor did he think the spiritual life was a matter of mere words, a construction of language, though of course from his point of view, the struggle to put it into words is more or less universally manifested in the world's religions, great and small. Hardy saw no reason to doubt that spirituality is part of the mainstream of our human experience, rooted substantially in our physical nature, what we are as biological organisms. He argued that the evolutionary process by which spiritual awareness emerged was of the behavioural type he had introduced in his first series of lectures. Behavioural selection is not blind, any more than the choice of birds to start pecking holes in milk bottle tops is blind. It operates through acts of conscious choice, that is, it is creative in the sense that animals, including the human species, have a hand in their own evolution. According to him, one might imagine that certain forerunners of *Homo sapiens* 'consciously chose' to attend to the spiritual dimension of their awareness. They did so because spiritual awareness gave them additional strength to cope with the dangers and difficulties of their physical and emotional environment.[36] Therefore, subsequent random genetic mutations that enhanced this kind of awareness would be selected for because they gave an advantage in the process of evolution.

What evidence did Hardy possess for this idea, apart from reflecting on the nature of his own spiritual experience? Much of the *a priori* support came from social anthropology. Anthropologists have often taken an interest in religion as a social institution, but not so many have considered the *experience* that Hardy assumed to underlie those institutions. One of the few who did was the anthropologist Robert Marrett. Here is Marrett gathering together his thoughts on a lifetime's work:

> It is the common experience of man that he can draw on a power that makes for, and in its most typical forms wills, righteousness, the sole condition being that a certain fear, a certain shyness and humility, accompany the effort so to do. That such a universal belief exists amongst all mankind, and that it is no less universally helpful in the highest degree, is the abiding impression left on my mind by the study of religion in its historico-scientific aspect.[37]

The pioneering French social scientist Émile Durkheim is often seen as explaining away religion reductively. However, Hardy felt that he had been misunderstood and quoted the following passage from Durkheim's great work, *The Elementary Forms of the Religious Life*:

> The believer who has communicated with his God is not merely a man who sees new truths of which the unbeliever is ignorant; he is a man who is stronger. He feels within him more force, either to endure the trials of existence, or to conquer them.[38]

Durkheim has a lot to say about this religious force, which he saw as generated by the collective consciousness of society as a whole. The term he uses for the physical feeling created is 'effervescence', the bubbling excitement that occurs in large religious gatherings (he had in mind the ritual corroborees of Australian aborigines). Durkheim's language is not always as clear as it might be, but Hardy was convinced that he did not take the next step of saying that religious experience is 'nothing but' effervescence, quoting in evidence the following passage:

> ... it is necessary to avoid seeing in this theory of religion a simple restatement of historical materialism: that would be to misunderstand our thought to an extreme degree. In showing that religion is something essentially social, we do not mean to say that it confines itself to translating into another language the material forms of society and its immediate vital necessities ...

> collective consciousness is something more than a mere epiphenomenon of its morphological basis, just as individual conciousness is something more than a simple efflorescence of the nervous system ... [39]

Hardy was a naturalist and like Durkheim he chose not to speculate about the nature of God, but to look at spirituality from the human side, the only way open to him as an empirical scientist. It is important to understand that he did not see the necessity to stick to scientific data as a restriction; indeed he criticised reductionist conjectures on the nature of religious experience precisely for their lack of a basis in empirical evidence. The embodied, sensual nature of human experience was what mattered to him, but at the same time in his private life he was an unorthodox but convinced and practising Christian, who attended Sunday services regularly in the Unitarian chapel of Manchester College (now Harris-Manchester College) in Oxford.

Although what he had to say about the evolution of spirituality was novel, Hardy recognised that there was a family affinity between his ideas and those of a number of previous students of religious experience.[40] Without exception these scholars came from a Puritan or Pietist Protestant background. That is to say, there is more than a suspicion that a particular form of religious education shaped their language and expectations about the experience of transcendence. No doubt there is a lot of truth in this, but there are surprising parallels beyond the evangelical Protestant fold that imply a more fundamental human basis. During the 1930s the German Benedictine monk Anselm Stolz[41] quite independently detected (and disliked) what he called a 'psychologising' tendency in Catholic theology, for example in the highly influential writings of the Jesuits Auguste Poulain[42] and Joseph Maréchal.[43]

Hardy of course was looking far beyond Christianity, for he believed that spiritual awareness is a human universal. For that reason, the sources he drew upon in his second series of Gifford lectures were extraordinarily wide ranging. Apart from anthropologists like Durkheim and Marrett (along with Bronislaw Malinowski, Godfrey Leinhardt and E. E. Evans-Pritchard) he also included what he called 'naturalists of religious experience', meaning specifically William James, Edwin Starbuck and Starbuck's Swiss/American rival,

James Leuba.[44] Among investigators of the numinous, along with Rudolf Otto he included poets like Wordsworth, Coleridge and Emerson, the mystic Richard Jefferies, the psychologist Cyril Burt and the novelist Marghanita Laski. He also referred to the work of students of animal behaviour, especially Konrad Lorenz and, more controversially, the reflections of H. H. Price and C. D. Broad on parapsychology.[45] Though he always talked of 'religious experience', his selection of interested scholars included agnostics and some who were positively atheistic (e.g. Leuba, Jefferies and Laski). This assortment underlined his belief that he was investigating a natural phenomenon that exists far beyond the world of formal religion and religious language.

COMMON CORE VERSUS COMMON CONTEXT

Hardy's claim for the universality of religious experience puts him close to the so-called 'common core' argument. One exponent of this view was Aldous Huxley (who incidentally was the brother of Julian, Hardy's old tutor at Oxford) in his book *The Perennial Philosophy*,[46] published in 1946. It is an anthology of quotations from a multitude of religious sources and is one of the best-known popular presentations of the opinion that, at the end of the day, all religions arise as responses to the same experiential core. Specialist scholars have frequently attacked the common core idea on the grounds that a detailed study of specific religions makes it implausible. Descriptions of religious experiences in different cultures are often utterly contradictory at the logical level, so surely cannot be referring to the same thing.

One carefully argued example of this kind of critique comes from the philosopher of religion Steven Katz, who invites us to compare Jewish and Buddhist experience.[47] On the basis of the *Torah* Jews are taught to think of God as personal, judging, and over against creation. Hence Jewish mystics experience God that way. They seldom if ever have moments when all distinctions are lost in ecstatic unity. Buddhists on the other hand are taught that there is no God and no Self. For them the approach to ultimate reality therefore involves, as a matter of course, the experience of the loss of self. Katz asks how there could possibly be any connection between such utterly different kinds of experience.

The contrast certainly is extreme, but Katz's reasoning does not necessitate the abandonment of Hardy's view.[48] Much of the scholarly analysis of experience in different cultures is done by examining the language of sacred texts like the Bible, the Qur'an or the Buddhist sutras. Scholars internal to the religion constantly comment upon such texts in their attempts to reach a clear understanding of what it is to be orthodox. Frequently their precepts are applied with great rigidity, leading to the expulsion or persecution of heretics. Their constrained accounts of what the devout person 'ought' to experience do not necessarily correspond with the varied dynamics of the life of any real individual.

From the opposite perspective, the way we understand metaphors nowadays is to see them as constructors of meaning. We also know that a very wide range of linguistic or symbolic expression can appear in association with what are certainly experiences common to every human being. Consider for example the incredible range of practice and interpretation surrounding the common act of eating food. Food that is highly prized in one community, for example pork, is considered disgusting and forbidden in other communities. These are culturally learned opposing responses, but no one disputes the need for food. Similarly, it is perfectly conceivable that different cultural descriptions of encounter with transcendence, or what the anthropologist Clifford Geertz calls the 'pristine powers'[49] could be utterly contradictory at the logical level. Yet this does not necessarily lead us to dismiss the existence of a more primordial common source. We know from the increasing amount of inter-religious encounter in recent years that religious 'professionals' from widely different cultures do seem to find common ground in the methodologies used to enter into contemplation or meditation. Many empirical studies have shown that they are in a similar or identical physiological state (this is a theme that will be returned to in more detail in Chapter 8). Even when the investigation moves to the level of conversation, accounts of meetings between Christian and Buddhist monks demonstrate the recognition of common ground.[50] Such dialogue is between people at least as culturally remote as Steven Katz's hypothetical Jewish and Buddhist mystics.

To suppose that 'religious experience' is a creation of the language and philosophical assumptions of a particular culture is to place it in a realm of extreme subjectivity. When you talk with people who have actually had the experience, it simply does not fit with this supposition and does not resemble the category of subjective production or imaginings. In crucial respects it is much more like the perception of an objective reality. I disagree with Katz, but I accept that there is something not quite right about the common core argument. People hold to religious beliefs for many reasons other than their direct personal experience. With this enlarged understanding, it is more appropriate to think of spiritual awareness as logically distinct from religion, and as the *biological context* in which religion can arise (but does not necessarily do so) rather than the common core of religion. A more appropriate way of referring to the biological precursor might be to call it the 'common context'.[51]

Hardy does say explicitly that he is talking about *religious* experience, but the biological nature of his argument means that, whether he realised it or not, he was implicitly going beyond the idea of an experiential core to 'religion' as such. According to him every member of the species *Homo sapiens* has this competence, hence this must include those who utterly reject religious belief. Accordingly, I conclude that Hardy was unnecessarily restricting himself by using the limited term 'religious experience' to refer to the subject of his investigations. For this reason, in my subsequent writing I will normally try to replace the adjective 'religious' with 'spiritual' to refer to the complete spectrum of such awareness, whether given a religious label or not. However, I shall quite often be forced by the context to talk of 'religion' or 'religious experience' because that is how an author chooses to describe their subject. It is important to bear in mind that when I use such restricted terminology I will have in mind that, from my perspective, it is a subset of the larger category of 'spiritual experience'.

TESTING THE RESILIENCE OF THE BIOLOGICAL HYPOTHESIS

Although he was able to identify numerous precursors of his ideas, Hardy's Aberdeen lectures are revolutionary in that they offer a testable naturalistic hypothesis about the nature and function of spirituality that is not reductionist in intention. In his inaugural lecture given in 1942 as Regius Professor of Natural History in Aberdeen University, he made his views crystal clear:

> I believe that the only true science of politics is that of human

> ecology – a quantitative science which will take in not only the economic and nutritional needs of man, but one which will include his emotional side as well, including the recognition of his spiritual as well as his physical behaviour. I believe that the dogmatic assertions of the mechanistic biologists, put forward with such confidence as if they were the voice of true science, where they are in reality the blind acceptance of an unproven hypothesis, are as damaging to the peace of mind of humanity as was the belief in everyday miracles in the middle ages.

By saying this, Hardy put himself at loggerheads with the major explanations of spiritual experience that dominate the universities to this day. I am thinking here in the first place of the conjectures of Marx and Freud which (at least in their origins) were attempts to account for the phenomenon of religion that they had previously decided on other grounds to be a universal human error. In spite of Hardy's remarks that I mentioned earlier, I am personally inclined to include at least some Durkheimian ideas as a kind of third member of a reductionist trio. I find his writing ambiguous on this matter, so possibly he himself should not be included. Nevertheless, many of his modern followers do seem to interpret his statement that 'religious experience *is* the effervescence or excitement experienced in crowded religious gatherings' as if it has the shadowy prefix 'nothing but'.

So how does Hardy's idea stand up to its eminent rivals? Since the foundation of the Religious Experience Research Unit in 1969 there has been a subsequent programme of scientific research exploring his thesis[52] along with parallel work in other centres in the United States, Australia and Italy.[53] In general the findings suggest that Hardy's hypothesis stands up to empirical testing better than its opponents. Marx, Freud and Durkheim are all highly respected icons of our culture and their ideas on religion have been the subject of very extensive study. But the great majority of these investigations look at religion as a social phenomenon. For example, research publications are often in the form of reports on how social and psychological variables relate to membership of a religious institution or to the holding of particular beliefs. Hardy was specifically concerned with personal spiritual experience. Here in summary are the scientific findings:

- People belonging to the poorest and most oppressed sectors of society, at least in the Western populations so far studied[54] are on the whole much less likely than others to report having spiritual experience. This is not what would be predicted from a straightforward interpretation of Marx's famous 'opium of the people' account of religion. If Marx was correct, it ought to follow that poor and oppressed people are especially vulnerable to hallucinatory religious experiences. The false consciousness of a non-existent father God who cares for them would act like a narcotic to dull the pain they endure in class-ridden society. All recent scientific data point overwhelmingly in the other direction. From Hardy's perspective the findings can reasonably be equated with the idea that poverty and oppression damage not merely physical health and happiness but also people's natural spiritual sensitivity.

  I need to add that I do not deny the validity of the claim that religious forms of fantasising can be used as a psychological defence.[55] Marx's loyal political partner Friedrich Engels explicitly condemned the religious devotion of the poor craft workers in the Wuppertal district of Germany. He saw it as a delusion leading them to be falsely happy with their miserable lot.[56] He may have been right. The institutional religion of Marx's Germany was certainly manipulated by the political establishment as a means of social control.[57] But the modern data suggest that Marx's critique is much too sweeping and over-influenced by his prior personal hostility to religion.
- The traditional (but not necessarily current) psychoanalytic assumption is that religious experience is a symptom of neurosis,[58] or temporary psychosis.[59] Taken at face value this leads to the prediction that in a large sample of the general population, people reporting such experience will score less well than others on a measure of psychological health. Using a measure developed at the National Opinion Research Center in Chicago,[60] the sociologist Andrew Greeley[61] showed that the reverse is true. In a national survey he directed in the United States in 1975, people claiming spiritual experience scored more highly on average on psychological well-being than others. His high scorers on mystical experience also got the highest score on psychological wellbeing of any group ever measured. In 1978 Ann Morisy and I[62] published the results of a repeat of Greeley's study for a British

national sample which showed the same result. This is what would be expected if, as Hardy believed, spiritual awareness is a necessary part of our biological make-up.

Unfortunately the waters have been muddied by Freud's frequently repeated claim, for example in *The Future of an Illusion*,[63] that people who suffer from the universal neurosis (religion) are less likely than others to undergo individual neuroses. But this is a perverse argument. It implies that those reporting spiritual experience, who in every other way are personally and socially competent, are in fact neurotics suffering from a temporary psychosis. To maintain cogency, the argument depends on a prior belief that religion is nonsensical and therefore such experience is at the very least illusory. Freud's argument is circular. The logic of his interpretation is not in dispute as it applies to certain kinds of religious belief and behaviour,[64] particularly its authoritarian forms.[65] But like Marx, he is trapped by his theoretical suppositions into making judgements that are too sweeping and not borne out by the empirical evidence.

- According to Durkheim, the primary source of religious experience is the assembly of believers. It is here that the effervescent representation of society is most strongly felt. Any such feelings experienced in private are secondary. This interpretation has been applied to other non-religious gatherings like pop concerts or football matches, and it is clear that in these cases the excitement of 'being there in the crowd' is the dominant issue. Applying the same logic to religious experience leads to the prediction that it will be less likely to take place when a person is alone. Ann Morisy and I published a paper on this question in 1985. We asked a random sample of adults in the city of Nottingham about the circumstances in which their experience took place. Approximately 70 per cent of those we spoke to said they were entirely alone at the time. This suggests that at the very least, Durkheim's conjecture is an incomplete account of such experience.[66] On the other hand the data are not incompatible with Hardy's hypothesis, especially in a society where religion is the object of scorn. Privacy gives protection.

A CRITICISM OF THE CRITICS

In spite of much scholarly criticism, Marx, Freud and Durkheim remain central to our contemporary understanding of ourselves. All three hold an honoured place in the academic curriculum and rightly so. They are major representatives of the mainstream of thought that has come down to us from the Enlightenment and I, like everybody else, remain indebted to them. To argue with them over one of their fundamental axioms is to get into a dispute with an entire culture and that is a daunting undertaking.

Nevertheless, I believe the way they went about their criticism of the experiential dimension of religion is, from a scientific perspective, methodologically wrong. Having been convinced by the eighteenth-century philosophical critique of religious belief, they were left to answer the question, 'Why in that case is religion so widespread, perhaps universal in the human species?' Their prior theoretical commitment, plus at times a large dose of distaste for the religious institutions they could see around them, meant that one of their axioms already in place was that religion is an error.

Although of course empirical science cannot avoid being theory-laden, it attempts as far as possible to start from the other end, by looking at the data. When we do that in relation not to religion, but the experiential dimension underpinning religion, it turns out that none of these reductionist explanations stands up to scrutiny. That does not of course mean that the diagnosis of religion as an error is itself a mistake. Nevertheless the data do make the sceptical critique of religious experience, and what I am now calling spiritual awareness, more doubtful.

Alister Hardy's originality grew precisely out of a stubborn empiricism that refused to bow down before a philosophical assumption. He would not allow his own experience to be denied or reduced and in this respect I believe he was more open-minded and less caught up in the presuppositions of post-Enlightenment culture than many of his eminent predecessors. Paradoxically of course, in breaking free he created a new kind of natural theology by applying the methods of empirical science that were themselves a major product of the Enlightenment.

This discussion has pointed towards one way of clarifying the confusion over the meaning of spirituality. Its close historical association with religion, even the notion that the two are synonymous, was in a way no mistake. Spiritual awareness is commonly the context out of which religion grows. But spirituality is not religion. Like Hardy I believe it is prior to religion and is a built-in, biologically structured dimension of the lives of all members of the human species. Therefore there are secular as well as religious expressions of spirituality, and many of them. Having set my perspective it is now time to have a look at the stories ordinary people tell as they try to make sense of their spiritual lives.

Part 2

CONVERSATION

Chapter 3

THE INDIVIDUALITY OF THE SPIRIT

> *The wind blows wherever it pleases; you hear its sound, but you cannot tell where it has come from or where it is going. That is how it is with all who are born of the Spirit.*
>
> John 3:8

The biological theory of spirituality (with its claim that everybody has it) leads us to expect that spiritual experience, especially these days, will sometimes be expressed in secular terms. In spite of that, religious language is still the commonly preferred choice, which ought to be helpful to researchers. It is easy to recognise and to connect it with related doctrines within the religious culture. Unfortunately standard religious statements can also operate as 'Keep Out' signs. A few years ago a young chemistry graduate who was a member of a fundamentalist evangelical sect gave me the following account of his conversion:

> Over approximately two years, both through the reading of the Bible and through the faithful preaching of the same, I came under a conviction of my own sin and was brought to faith in the Lord Jesus Christ. It's a God-given thing – not at all the work of man.

I was keen to know what was special in this young man's case; the subtle facets of his personal feelings at the time of his conversion. He mentioned none of these. Perhaps they were beyond words. Perhaps he felt the particulars of his conversion were none of my business. My impression at the time was that he was conforming to an expected pattern, and that in doing so, his terminology concealed as much as it revealed. I could almost repeat by heart the formula he used, for I was familiar with it from the religious environment of my childhood.[1]

There have been many times and places where the stories people tell about their spiritual experience are pretty formulaic. This is particularly likely to be the case where a religious institution is socially dominant. In traditionally devout communities, people are taught what is supposed to happen as they advance in the spiritual life, sometimes in considerable detail. Theologians have laid out the pattern for them, possibly hundreds of years previously, giving it immense authority as timeless truth. The old *ordo salutis* of the Lutheran Church had ten stages that believers were expected to go through. First they were elected, then successively called, illumined, converted, regenerated, justified, united mystically with Christ, renovated, preserved to the end, and finally glorified with the Son; a complex, not to say wearisome sequence.[2] Within the Roman Catholic tradition, textbooks on prayer are still available containing very detailed, step-by-step accounts of how to progress in prayer. Adolphe Tanquerey's *The Spiritual Life*, published in 1930 and at one time to be found in seminaries and convents throughout the Western world, is a classic example of the genre.[3] It advances sequentially through the spiritual life and reads almost like a technical manual, subdivided into more than 1,500 numbered sections.[4]

TALKING TO ORDINARY PEOPLE

It is highly unlikely that such standardisation would be found in the modern West, except amongst the membership of a few relatively isolated religious communities. In that case, what about the situation of people who never go to church and don't necessarily subscribe to any standard religious formula? If the biological thesis is right, they certainly ought to have a spiritual life. My colleague Kate Hunt and I decided to try to find out. As the number of regular churchgoers in the UK is now very small, we didn't anticipate any difficulty in identifying people who fitted the bill. In consultation with Gordon Heald, we wrote a questionnaire on areas like religious identity, church attendance, belief in God, self-identity (as either religious, spiritual, agnostic, atheist etc.), spiritual experiences, and attitudes towards the meaning of life. A pollster employed by ORB stopped people randomly in a Nottingham shopping centre and asked if they would respond to the questionnaire. If they answered a question about churchgoing by saying they never went to church, they were candidates for selection for interview.

To be on the safe side, we felt it necessary to add the requirement that they labelled themselves 'religious' or 'spiritual'. This was potentially a serious error because it could mean that the people chosen were a small subset of the non-churchgoing population who happened to take an interest in 'that sort of thing'. Fortunately we were saved to some degree from our folly by the statistics from our national survey (see Chapter 1) which told us that three out of every four of the people who passed our pollster in the shopping centre would probably claim to have a spiritual dimension to their experience. Nevertheless the resulting limitation to the data needs to be remembered. We invited a sample of those who fitted our criteria to attend one of four focus groups held a few days after the initial interview.

Although we discussed a large number of possible ways of making up the focus groups, we decided to concentrate on differences in age and gender (see table in notes).[5] The first group consisted of men and women within the age range of 20 to 39, whilst the second group was also mixed gender, within the age range of 40 and upwards. These two groups were set up on the intuition that the decade of the 1960s was a pivotal period of cultural change.[6] People born before then are more likely to have had some sort of formal contact with the church than those born afterwards. The third focus group was made up of women and the final group was all male. We divided these latter groups along gender lines to test the assumption that women are more concerned about spiritual issues than men. At the conclusion of the focus groups, we asked each of the participants if they would be willing to be interviewed on their own by one of the researchers. Only one person declined to continue.[7]

In general we found that to varying degrees people were defensive at the beginning of the conversations. They gradually relaxed as they realised that we did not have a hidden agenda either to criticise or to convert them, and that we were genuinely interested in their own views. Although we were researchers we were perfectly well aware that we were not totally detached outsiders, but fellow human beings with our own life stories and spirituality. These would inevitably affect the mood of the conversation and the best we could do was to be alert to these effects. Our aim was to discover people's personal understanding of their individual spiritualities. Because of this, we did not define the terms 'spiritual' or 'spirituality' at any point in the research process. This follows from our stance that spirituality is present in some form in everybody's life and is not necessarily connected to religious beliefs and practices. As a contemporary thinker on spirituality writes:

> The spiritual story itself is much more powerful and coherent than any text-book definition or description of spirituality.[8]

Therefore we were interested in people's feelings and life experiences, and not just in their intellectual beliefs. We found that as a conversation progressed it almost automatically took on recognisably spiritual dimensions as the person wrestled to articulate their own sense of the mystery of their existence. Many admitted that they had not had such an opportunity before and found that the experience of talking to us forced them to look at their own lives in a different way. This made us very sensitive to the power of our own role as researchers, and the need for respect towards the people with whom we spoke.

In the arena of the focus groups there were many signs of reticence in speaking about spirituality, where assumed norms of what is acceptable in public were operating. Concealment of this kind was hardly ever a problem with the people we spoke to in our private research conversations. Most talked very openly about the many twists and turns of their spiritual journey, quite often explaining how they had extricated themselves without regret from the religion of their childhood. It was as if they wanted to emphasise that they were individuals who had broken away from a constricting pattern. Even so, it was striking to see how often they continued to use orthodox religious language – and this included people who were very remote indeed from the church. It seemed that half-remembered phrases had been hanging around in a lumber room at the back of their minds, waiting for a moment when they could be used again legitimately.

NICOLA, MATTHEW, TOM AND STEPHANIE TELL THEIR STORIES

I would now like to share something of the varied flavour and individuality of people's spiritual narratives. As illustrations, I'm going to let four very different people with whom we conversed tell their story. (The names of the interviewees have been changed throughout.)

'IT'S STILL THERE'

Nicola was a lively and talkative woman in her late forties, born in England of an Italian mother and a Scottish father. Her mother was, in her eighties, still a daily Mass attender and Nicola was obviously very fond of her, talking movingly of the small community of elderly Italian ladies with whom her mother socialised at the church. Not surprisingly, Catholicism had been part of Nicola's cultural identity all her life. She had stopped going to Mass at the age of sixteen when she left home. Nonetheless, she still had more than a merely tribal allegiance to the faith of her childhood, perhaps strengthened by her perception of her mother as a very good person.

The focus group gave Nicola a rare opportunity to spend time reflecting on her hitherto somewhat submerged beliefs. She was quite happy to admit to the group that she was a Catholic, but also let it be known early on in the conversation that religious observance did not play any part in her adult life. As the evening wore on, much to her own amazement, Nicola found herself taking on the role of 'defender of the faith' against the attacks of thc other members of the all-women focus group. The discussion had quickly turned into a severe critique of religious belief in general and more specifically of the institutional church. When someone remarked 'Look at all the trouble over the world to do with religion', Nicola snapped back, 'That's an excuse. I don't believe that … I think religion is just an excuse they use. It's just the type of people they are.' One member of the group who worked with terminally ill people raised the familiar conundrum, 'If there is a God, why are people suffering?' Nicola's response revealed that at some point she had undertaken a sophisticated reflection on the problem:

> I suppose he suffered and if you believe the beliefs of the Bible, he didn't take himself away from the suffering. Things happen and it's not all in his hands.

At the beginning of the private research conversation, Nicola commented on how surprised she had been by her behaviour in the focus group and also by the strength of her convictions:

> … you don't sort of think about religion, and how you feel, as such. You don't sort of sit there and think, 'Oh I believe in this, and I believe … ' until somebody is talking to you, or questions you about it, or it comes up in conversation, and it comes out, 'Oh crikey', you know [laughter] 'it's still there!'

As Nicola reflected on her life during the discussion, she discovered that her beliefs were still very strong. But this only happened because an environment had been specifically created for her to do this. Most of the time Nicola did not allow herself to think particularly deeply about life, although as we have seen she was able to draw upon subtle reflections on difficult religious matters. It was as if the surrounding culture had imposed a block on sustained investigation of the spiritual dimension of life.

> I think you just have to make the most of what you've got at the time, and not think too deeply about it. Because I think if you go into it too deeply, and think about it too deeply, I dunno, you can sort of, 'Why am I here? What am I doing here?', and it could be sort of a bit negative feeling, if you aren't coming up with the right answers.

Nicola acknowledged the importance of her Catholic heritage, but preferred to keep busy rather than give herself permission to face the big questions of existence. Although she had no current links with the institutional church, she had tangible proof of her continuing status as a member of the church in the form of various religious objects that she kept in a box in her attic. This 'reliquary' contained her First Communion dress, a Missal inlaid with mother of pearl and a children's Bible. These items were not taken out of the box, but they had power just by being there; they proved that she still belongs.

Later on in the conversation Nicola talked about her belief in God and her Catholic identity as in reserve, ready to be used in emergency. Her reliquary perhaps held the same function, tying her to her roots in the Church community.

> I know it sounds stupid, but it's like having some money in the bank, at the back of you, if you ever need it. Not sort of the money that you can go into every day, but just some security

> there at the back of you, in times of crisis, or if something goes wrong, you've got something there. And I think I feel like that about being a Catholic, you've got something there, you've got your belief, there's not nothing there. I know it sounds ...

The key phrase here is 'there's not nothing there'. Nicola's faith provides her with a way out of existential loneliness. She is part of something greater than herself, the Catholic Church, which provides her with the security she needs to cope with life.

Nicola went on to describe a time in her life when she needed to raid her deposit box:

> I think the worst thing was when my husband had his heart attack. Um, that's a horrible experience, I would hate to go through something like that again ... I think that's sometimes when you feel an inner strength, and there's a presence there to look after you.

In Nicola's mind, she still belongs to the institutional church; she is an insider, even though she does not attend.

> I still feel I've got a sort of a connection with the Church, with my Catholic faith. I don't think that will ever go. And you can either hate it, or you can carry on with it like Mum, but I think deep down inside, it's always there with you.

This is a recurrent theme in all those research conversations we had with people who had some kind of sustained affiliation to the church as children. They talked of 'the seed having being sown' and of a sense that they cannot escape either God or religious belief; it is part of their inheritance. Nicola pondered whether she believed simply because she had been indoctrinated, but decided that it was more than that. Her life's experience confirmed for her that her belief was genuine. This is where a contrast appears compared with those people whose association with the church was at best tenuous. Matthew, Tom and Stephanie, who are discussed below, either do not have access to, or find implausible, traditional religious language and symbolic frameworks with which to express their spiritual beliefs and experiences. Others, such as Nicola, are still able to make use of the resources of the Christian culture.

If Nicola's identity as a Catholic was still so important to her, then why didn't she attend Mass? Unlike many others, Nicola was not angry with the institution; far from it. Why was her membership passive rather than active? When asked why she didn't go to church, Nicola came up with the most familiar of all reasons; she had no time.

The problem is that time has become a commodity. Time is money. The industrial revolution completely changed our attitude towards it and made it one of the most precious commodities of all. For most people, communal religious practice is definitely low down on the list; it is not a good use of time. This is usually interpreted as indicative of a weakening of the power of religion. It may also in part be plausibly interpreted as evidence for the privatisation of religion. People may ask themselves what need they have of the larger community when they enter the deepest part of themselves. This discarding of community goes along with the extreme individualist characteristic of modernity that I discuss in a later chapter.

Nicola remarked, 'I've always liked to do my own thing, and not feel tied, which is going away from the Catholic belief, I know.' She sees going to Mass on a Sunday as unnecessary, but in spite of that her mother acts as a kind of proxy, and she knows that the church will still be there when she needs it. There is a tension between Nicola's traditional upbringing, which is still important to her, and the demands of contemporary life. Nicola wants to be part of the Catholic culture that she continues to admire in her mother; but it is incompatible with modern living. For people such as Nicola, the erosion of community is incomplete. She is caught between two worlds.

'WHAT IF, PAL?'

Nicola is someone who, in her detached way, still remains within the orbit of the Church. Matthew, on the other hand, definitely sees himself as an outsider, severely criticising what he perceives to be the church's claimed monopoly on spirituality. He is very gentle, very reflective, a professional man in his early forties. He is divorced and is able to see his children only at weekends. He has had much less connection with the religious institution than Nicola. What he remembered in the focus group was one year of Sunday School at a Methodist church when he was nine, and the obligatory religious assemblies at the beginning of morning school.

Matthew's conversation illustrated very clearly the attitude towards institutional religion of many of the people we spoke with. Unlike Nicola, he felt unable to accept the over-arching Christian account of ultimate reality. He was inclined to a pantheistic belief, that is, the idea that the universe *is* God. Nevertheless his conversation continued to contain fragments of Christian discourse. Traditional religious language was to hand and he used it when he needed it, whilst at the same time being sceptical of its validity.

Matthew began the private research conversation in contradictory fashion. He admitted that the whole subject of spirituality was 'immensely fascinating', following this with a declaration that there is no point in trying to search for meaning and purpose in this life:

> I don't think, you know, the sort of 'why' question is really relevant … I don't feel the need to keep, you know, this, you know, 'why are we here?'. We are, it's marvellous, but we are, and it's a fantastic gift, but I don't think we should be wasting, or concentrating too much time, struggling and grappling with it.

In fact, he admitted that he couldn't help struggling and grappling with questions of meaning. There was something in him that wouldn't let him stop, as became clear when he spent the rest of the conversation attempting to put into words his own search for a coherent belief. An unnerving contradiction lies at the heart of Matthew's spirituality; the overpowering desire for some kind of certainty along with a recognition that, as far as he is concerned, this is impossible.

The apparent certainties of some religious people got on his nerves. He criticised what he called the 'ring-fencing' of beliefs by particular religious groups (Christians, Muslims) as a way of setting oneself apart from people outside the fence, making the outsiders seem less human. He made a heartfelt cry for toleration whilst recognising the human longing for certainty:

> … you can't embrace everything obviously, but you can find … toleration possibly, that you allow others the freedom to believe whatever they wish to believe … and entertain the possibility that what you believe can at any point be challenged, but is nevertheless important, and viable and truthful, to yourself … because you see, I think that we all of us need this, you know,

> we need to cling on to certain private certainties, and … certainties that are shared by a group that we wish to associate with … whenever that happens there's always going to be the tension because … by reinforcing the beliefs within that group, you're distancing yourself [from all other groups].

Later he said heatedly,

> … why do we have to be so bloody petty, and squabble about stupid, fantastic, supernatural nonsense? And you know, whilst I say that, I still have this belief in something … I don't believe he's this shape, and he wears robes, and he's got a beard … I've seen the way the Church of Latter Day Saints depicts Jesus … and he's this square-jawed Aryan, you know, fantastically handsome … I do find it amusing, but it's more than that, in fact it's pernicious … What about the hunch-back, mixed-coloured guy, you know, and he sees this? And it's élitist, it's sentimental, you're just excluding people … I realise we are probably required, we have to put a shape and face to him, I suppose … it's not on to just be happy with something amorphous …

Behind these vivid and contradictory feelings, Matthew mentioned three distinct influences on his life, two of which told him that life is chaotic and meaningless and one that left open the possibility of meaning. The first influence was his personal history. Two critical incidents had impacted on him profoundly; the death of a school friend from leukaemia when he was eighteen, and his recent marriage break-up. These events made him realise that, as he says,

> … this isn't adding up to going any place, it's just, again it's such arbitrary, chaotic tragedy and nonsense. Yeh, I think that's when the whole, you know, meaninglessness of things, um, really started taking some sort of definite shape.

Matthew had looked to the Christian tradition for some explanation of these tragedies, but found no solace; for him there is no supernatural person caring for us: 'No, it doesn't work like that, we don't get looked after on that individual level.' As far as he could see, the church still portrays God as the old man in the sky, an idea that to him is 'self-indulgent', 'awfully sentimental', and 'just wildly wrong'. God does not intervene in his life, or anyone else's; he is at best an 'absentee landlord'. This is not the kind of God that Matthew can believe in.

The second influence on Matthew's life was the world of science. If Christianity, or any formal religion for that matter, could not offer him an explanation for life, then perhaps science could come to the rescue. He found some comfort in the idea of evolution and of human progress. But even that did not ultimately help him, as he knew only too well that so-called 'progress' can be used for evil purposes as well as good. Science could tell Matthew the awe-inspiring fact that the atoms in his body were once part of a star, but it did not answer his deepest questions of meaning.

The third influence was the 'nagging instinct' that told Matthew that there must be more to life than mere existence:

> ... it's probably just a nagging instinct, that while all the material evidence is telling me, this is ludicrous, you know, this is all complete chaos, nonsense, it's arbitrary, you know, we're a rock in a vacuum just spinning through nothingness and, you know, the consequence of impersonal cosmic forces, nothing beyond it. Whilst my sort of intellectual faculty can tell me that, there is this other, and I'm not going to use the word 'soul', but there's this other bit of me which is just sort of going, 'hang on', you know, 'What if, pal?'

It is the 'what if?' that keeps Matthew on his spiritual search.

In our culture, the obvious place to start this kind of search has traditionally been the Christian church, and that is where Matthew started, but the institution had been a big disappointment. It is important to note that his view of the church was not based only on a dislike of popular images of God. He also felt very strongly that religious people in general are arrogant in their assumption that they are the bearers of the truth:

> ... it seems so very often that you ... people that are believers, they've got it, they know. The fact that you don't know means that you, yes, you don't know, you're stupid. It's the arrogance, um, it's a very well-dressed-up humble-looking arrogance, but it's arrogance nonetheless, however. That's what annoys me. You know, you're not with us, therefore you're against us. No, I'm with you, but I'm not sure. I'm not against you, and I just

> want to know. If you know, how do you know? Do you really know, all the time?

There is a poignant sense of yearning here, a desire for some kind of connection. The problem is that for Matthew, and many other people we spoke with, belonging to the religious institution means that there is no room for doubt, you have to believe with certainty. And this is just not possible for him as he reflects on his own life narrative and the world revealed to him by science. Matthew's spirituality is not dead; on the contrary it is dynamic and very much alive. It changes with his life and because of that there is no room for a once-and-for-all revelation.

Yet he gave the impression of longing to be able to belong to a faith community, a place where he could explore his beliefs and develop his spirituality. He wished the church was less dogmatic, but also that it was much more serious about its role instead of being bogged down, as he saw it, in empty ritualism and fine words. To him this seemed like a kind of frivolity when considering such a serious matter:

> But again, it's being an iconoclast, smashing that nonsense down, you know, get real, let's get underneath all that …

He adds somewhat wistfully,

> I think they get a lot out [of attending Church], again, this is probably envy in me: why don't they invent one that I can go to?

Unlike Nicola, it is not possible for Matthew to feel part of the church, to belong. He has no safety-deposit box set aside for a rainy day. He lives with the enigma of existence every day and finds little solace. 'Some days I think I know, and then I don't really.'

The conversation with Matthew vividly illustrates the struggle that many people have to nurture their own spirituality when deprived of either the language or the support of an institution because they find them implausible or unattractive. The church is out of bounds for the likes of Matthew, and so he is left to search in other places.

> And that's probably why I do tend to explore the subjects and the religions. I will talk to people, find out what they believe. Because maybe one day I'll find it, you know. But again, I think there is a need in all of us, I think; however much we resist it.

Although Matthew rejects Christianity, his own search for faith is in constant tension with the Christian tradition. It is almost as if he realises that the Church should offer him a place in which to share his spiritual quest, and he is angry that, as far as he can see, it does not recognise his search as legitimate. Quite suddenly, towards the end of the conversation, Matthew began to speak movingly and with deep emotion about his love for church buildings.

> I still love the church, you know, at Christmas time. I like going into churches, from an aesthetic [point of view], I think they're great places. They have weight and silence and tranquillity and beauty, and you can find solace.

'THERE'S ALWAYS MORE TO LIFE. THERE'S ALWAYS, NEVER ONE TRUTH'

Unlike Matthew, Tom was someone for whom institutional religion figures hardly at all. Tom was in his thirties and quite 'laddish' in appearance. He was married and worked in a small local firm. When he was a child he sang in the choir of his local parish church but that ended when he was fifteen. The only time that he could subsequently recall attending a church service was for his own wedding. The impact of his youthful association with the church choir seemed to be zero. Nevertheless he was quite clear that he was deeply spiritual:

> I believe more in Mother Nature than anything else. It's like an electric field that combines everything, whether it's plants, humans or anything else.

He had developed his own brand of spirituality from a mixture of folk religion, magic and popular science. His conversation illustrates the spiritual creativity of someone who is almost completely outside the mainstream Christian tradition.

Tom's spiritual life was of a highly materialistic sort. His talk was peppered with references to tangible phenomena such as ghosts, apparitions, premonitions, 'atmospheres' and the like. For him the spirit world is all around us if we care to look. He had been to a spiritualist meeting a few times and was quite impressed by the phenomena he had witnessed. For example, during a séance he had felt that one side of his body was on fire, 'really, really sore and hurting'. It turned out that a woman sitting near him had lost a son in a fire, burned down the same side of his body. Tom had also tried using a Ouija board, had investigated spontaneous human combustion and had 'a collection of books upstairs' on similar phenomena.

He was quite clear that there is no God, though he speculated that human beings might have souls. Apart from that, his metaphysical beliefs hardly went beyond the view that 'everything's got to be combined somewhere down the line'. His spirituality was thoroughly pragmatic. He explained that he and his wife Sharon sometimes took on the practical role of 'shamans' in their immediate circle, getting in touch with the beyond on behalf of their friends. Sharon was the seer, as Tom said in a matter-of-fact way:

> Right, as far as premonition's concerned, I mean that's, that's definitely Sharon's field and whenever she has one, I listen.

Tom enjoyed leading ghost hunts in the local woods. As he described these jaunts, he moved from treating it all as a joke to taking it very seriously, as he spoke of witnessing things he could not explain rationally. This ambivalence was present throughout his conversation. At times he was dismissive and laughed at his beliefs, in what felt like a defensive way. It seemed likely that Tom was reacting to the gendered cultural expectation that men have to be cynical and objective in these realms. It was only as he relaxed and got the feeling that he would not be listened to dismissively, that he was able to talk about what he really felt about his experiences.

Another example of Tom's spiritual practicality was his much-appreciated ability to feel atmospheres in people's houses. He explained that his friends sometimes asked him to go to their new homes to judge whether or not the place was friendly. So the spirit world truly is part of Tom's everyday life. But again, he was well aware of the cultural critique of such beliefs. When referring to the friends who ask him to check out their houses, he said:

> And they, they wouldn't come out directly and say, 'I want you to tell me if you sense anything in the house.' They'll say, 'just come and have a quick look at my house for me.' So they're not so much admitting to themselves what they are doing as sitting on the fence. They'd rather hide it and not say anything in public, 'cos they might be criticised for what they say.

Apart from his initial defensive ambivalence, Tom was perfectly happy to talk about his psychic experiences, perhaps because he was brought up in a household where such events were considered the norm. Tom's grandmother 'used to delve into all this sort of stuff as well, bless her', with her copy of *Everybody's Fate and Fortune.* He also described how his sister fixes watches by staring at them. So, as Tom himself acknowledged, it was not something he had ever been far away from. His family thus gave him important social support for his beliefs.

Towards the end of the conversation there was a change in tone as Tom began to talk about a very different kind of experience that he had eight years previously. His father had become seriously ill and Tom had to take him to hospital in the middle of the night. He described the scene:

> As we sat in the waiting room, you know, my mum was absolutely distraught, and I just sat there praying and praying. Not shouting to anybody, but just sort of 'if anybody can hear me please help'. And it was really weird and I don't know if it was in my own mind ... but I just suddenly felt really warm and it felt like somebody had just turned around and said 'Don't worry, it's going to be OK.'

This episode has the hallmarks of very many hundreds of the accounts of religious experience collected by Alister Hardy in Oxford. The crux of such experience is not some kind of supernatural reassurance that the fraught situation 'will be OK' in the material sense. Quite often nothing changes materially, or things may even get worse, but the circumstances become bearable, and there is a sense of being placed in a larger, ultimately benign context.

When Tom introduced this episode rather shyly, he called it 'another little incident in life'. Why did he choose to underplay what he obviously felt was a very dramatic moment? Perhaps we are back with the taboo about admitting to spiritual experience. It seems significant that he was most circumspect in mentioning the experience. He told his mother and subsequently his wife, but no one else until the research conversation. He was clear that he would definitely not have said anything about it at the focus group, ''cos I don't really admit it'. Why is that? What is the difference between giving accounts of ghost hunts and spontaneous human combustion, as Tom does freely, and talking about a religious experience?

The issue seems to be one of material proof. Tom was happy to talk about certain kinds of beliefs and experiences because in his mind he was able to provide tangible evidence that they 'really happened'. He was perfectly well aware that we live in a scientific age where physical proof that fits with the canons of the scientific method is highly prized. Tom tried hard to keep these two distinct parts of his experience together: his keen spiritual awareness and love of mystery, and his need for a secure basis in the proven physicality of the world. He was able to keep these two in tension with his highly materialistic understanding of spirituality, but it was not so easy with his experience of sitting praying in an empty hospital corridor.

Tangible proof is a theme running throughout Tom's conversation. He states that the reason he believes scientists is that they have facts and proof, and that he would even believe in a religion such as Christianity if they could show him some physical evidence.

> I don't mind the idea of believing in something, if it's there in my face and I know it's there. But I haven't been given any other, any, any factual evidence and I do go a lot on sort of facts.

No wonder he described his experience in the hospital as 'so bizarre' and 'really, really weird'.

Unlike Matthew, Tom did not focus on a critique of the institutional church or Christian doctrine. It was as if these were so far removed from his life that they did not figure in his reflections on spirituality at all. But there is a contradiction here once again, as he did admit to being interested in church buildings. He even said that since the focus group meeting he had found he wanted to go into churches as he walked past them.

> I've been past a couple of churches and just felt a need to go in. Although I'm not spiritual (sic). Which is sort of quite strange for me. I do get a little impulse now and then but I just shrug it off …

What is this impulse? It seems to be linked with Tom's interest in atmospheres and, once more, his desire for tangible evidence of the spiritual world. He tried to explain his reasons in this way:

> Maybe I want to believe. Maybe there's a sense of I want to

> believe in something definite and concrete and maybe I've got the idea that if I go in at that moment, I might see something.

Churches are not dead, empty shells for Tom, they have the potential to reveal spiritual truths. His experience is that the walls carry a reverberation of the beliefs and prayers of people throughout history, and he finds this appealing,

> Rock, brick – it's like a tape recorder. It echoes things from the past.

But there is never any sense that Tom wishes to belong to the present day community of faith. This is simply not an option for him. Religion is forever linked with the *Songs of Praise* programme on Sunday evening television, and the rush to see which member of the household can switch it off first. It has nothing to do with Tom's personal belief system.

Tom illustrates a tension which exists in nearly all the conversations: how to maintain one's integrity as a member of a supposedly highly rational, logical, scientific culture and yet at the same time allow one's spiritual awareness to flourish. His conversation may be confused and self-contradictory at times, but this is precisely because he is attempting to hold these two contrasting worldviews together. Tom is aware of the prohibition surrounding admitting to one's own spirituality, but seemed to value the opportunity to share his thoughts in the research conversation. After being asked why he was interested in spirituality, Tom replied:

> It's just a sense of, er, I don't know. It's a sense of wonder, isn't it? If everything's as black and white as everybody wants it to be painted, the world would be a dull and boring place. And, and it's the not knowing. It's not, it's like is there life on other planets? It's the not knowing. [It's the mystery?] Yeah. I think if something came down and presented itself to you then the mystery would be gone ... See, I'm not really weird [followed by laughter].

'WE'RE ALL HEADING FOR THE SAME PLACE'

In two respects Stephanie was rather different from almost all the other people involved in the research. Whereas most of them lived in the suburbs or in the country, she lived in an inner-city area of Nottingham. She was a single mother with two small children. As we spoke, it became clear that life had been tough for her and her personal history included suffering violence from a former partner. Spirituality played a pivotal role in helping her to cope with the special demands of her everyday life.

Stephanie's spirituality falls into the category of New Age thinking. She was unusually outspoken about her beliefs, whilst being well aware that for many people they were 'off the wall'. As she said of her presence in the focus group:

> I was sitting there quite a lot of the time and thinking, 'I don't think I'm what they expected really', do you know what I mean? ... I felt like, you know, a bit of a loose cannon in a way ... Like a bit odd, compared to everyone else.

During the conversation Stephanie spoke unselfconsciously of her belief in reincarnation, the power of meditation and the 'Universal Consciousness' that binds everything together. Yet Stephanie's spirituality is not simply esoteric; it is a serious business for her. This was clear from the way she spoke in the focus group, uninhibitedly and without the slight feeling of embarrassment felt by several other members of the group. Stephanie's spirituality in some respects was like Tom's in that it was a 'spirituality of practice':[9] she makes time to meditate, read books and connect with the universe. But, compared with Tom, she had an air of intensity, and her commitment pervaded the whole of life. So this is quite a different spirituality from, say, Nicola's. Instead of being a handy life-insurance policy or backdrop to life, it is centre stage, informing her daily living.

Spiritual awareness had been part of Stephanie's life since childhood. She described her youthful musings over what it means to be human:

> I've always been very aware of my own mortality, even when I was a little girl. You know, when you look in the mirror, and you see yourself, and you don't really pay any attention to it on a day-to-day basis. But I would stand, I'd look in the mirror and I'd be kind of, 'that's me in there'. Do you know what I mean? I was very aware of self, and the fact that, I don't know, but bodies seemed a bit inadequate in a way for everything that was

> inside. And I used to stand there, and I'd look at my eyes and like, trying to see what was inside.

But Stephanie is aware that our culture tends to disparage these kinds of thoughts.

> But then you are taught things at home, and at school, which take you away from that, which is a sad thing. And I kind of, I let it go I think for years. I was always still very aware of that, looking in the mirror and things like that but, like I say, you're taken away from it.

Stephanie had decided about three years before this conversation that she wanted to renew the spiritual inquisitiveness that she had experienced as a child looking in her mirror. She had not gone to the traditional religious institutions in her search because she found them too narrow and too obsessed with pain and suffering. Stephanie knew only too well what pain is, and indeed:

> You should feel it sometimes, because I believe like, you know, with pain, there's no way around it. You can't go over, you can't go under it, you can't nip [out], there isn't a side door, you've got to go right the way through it, feel it, and then you get to the other side …

But she went on to say:

> The sad thing about religion is like, you know, they say the easy path isn't the path to spiritual enlightenment or whatever. And you have to walk down this road with thorns and things like that. I don't believe that is necessarily the case. I believe the path to spiritual enlightenment can be beautiful.

She was in effect criticising the church for claiming (as she saw it) that true spirituality is only about pain and suffering. Her image of the church was of a long line of people following the leader, the priest, on a narrow path towards the gates of heaven. Stephanie's own view of the spiritual journey is one where people join hands together rather than being in crocodile file:

> This path that we go on, I don't believe that each path is a separate thing. I think there are points in the way where each path will meet, and you have a choice to say, 'Oh actually I'll

> join. Can I walk with you a while?' And that's the way I see, when I'm with certain friends who believe in the same kind of things, it's like, we'll walk together a while, you know, and enjoy the scenery together. We'll smell the flowers together, and that's nice. And then there are other times when you do it on your own, simple.

Stephanie's spirituality is therefore both communal and personal. But for Stephanie there is a need for freedom within community. She seems to be reacting against particular kinds of false community, the 'tribalism' of certain groups, where people hold together on the basis of some ideology (political, religious, sporting) for reasons of personal security. They see themselves as behind a barricade, defended from a hostile world. Stephanie's universalism is the reverse of this; she insisted on openness and individuality at the same time:

> I don't think any of us can do it alone, but we should be allowed to do it in our own way … the oneness and the wholeness is the same thing to me, if you see what I mean.

Stephanie went on to talk about the life-giving nature of her relationship with the 'Universal Consciousness':

> I spend a lot of time on my own, so I'm very used to being a single unit, as it were … And I have experienced vast chasms of empty loneliness. When the children have gone to bed, if I'm on my own, I have on occasions felt extremely alone, and very empty. And getting in touch with the Universal Consciousness is like being in touch with everyone else who's out there. And it's like filling yourself up with that, to take that feeling away.

Although Stephanie was adamant that her spirituality is separate from Christianity, there are obvious similarities. In the research conversation she often seemed to be talking, so to speak, in parallel with orthodoxy. In discussing beliefs and experiences that are very familiar in Christian rhetoric, Stephanie used her own words. A pivotal example of this was her personification of the universe as the 'Universal Consciousness'. For Stephanie, the universe is a real entity and she talked of it in similar terms to the Christian view of 'God'. She spoke of 'sending out a message to the universe', which sounds very like the idea of intercessory prayer. She expected the universe to respond to her messages:

> The universe is a wonderful, bountiful place and it will give you whatever you want, provided you send out a clear enough message.

Forgiveness was another very important part of Stephanie's spirituality. She spoke of the need to forgive those who have caused her pain and acknowledged that real forgiveness is extremely difficult. She even went on to use biblical language to explain her belief that revenge is unnecessary.

> What you put out, you'll get back tenfold. The universe will look after me. The universe will look after them. If they've done something bad to me, the universe will take care of them, not me, it's not my place. I mean, your God, as you call him, like 'vengeance is mine, sayeth the Lord,' well, it's that theory, but in my words, it's the Universe will take care of it, I don't.

Stephanie felt that it is important to send out 'unconditional love' both to the people she cares for and also those who have hurt her. These are all concepts familiar to Christianity, so why was Stephanie so adamant that she did not want to be associated with this particular institutional religion? Her reasons seem to be to do with authenticity. Stephanie's spirituality incorporates a keen sense of social justice and a desire for equality. She felt that the church fails to fulfil its duty to the poor and to the environment:

> The problem now is that religion is a deeply organised thing. And I think the church has got a very hard time convincing people of its motives when you look at things like, the Catholic Church is the richest thing in the world. Why is it not sharing that? How can it claim to be what it says it is, when it has got all of this wealth, and it is just sitting on it? As far as I'm concerned the Catholic Church is sitting on a pile of eggs that have gone off, because it's not doing them any good sitting in banks, it needs to be out there and helping people.

When we discussed the Bible, Stephanie argued that it needs to be interpreted in a way that makes sense to her at the end of the twentieth century. She acknowledged that there was wisdom in the Bible, but times have changed and Stephanie's needs and concerns are different to those of biblical times.

> I worry about my children going on school trips, and the bus crashing. I have those kinds of worries. Someone two thousand years ago might have those same worries related to something else. It's about bringing that so that that makes sense to me, in my life, reaching out to me today, not two thousand years ago.

For these reasons, Stephanie had chosen to find her own spiritual path. The echoes of Christian tradition still resonate surprisingly strongly, but she has created her own language and imagery to explain her spirituality.

SUMMING UP

I chose to discuss these four people because they show vividly contrasting responses to the spiritual search. Their perspectives are by no means exhaustive, as will become clear elsewhere in this book. But their range is striking. Nicola is quite clearly emotionally the closest to the Christian institution. Her spirituality is still very much expressed in traditional symbolism, even if that symbolism is currently stacked away in the loft. I have the feeling that had she been an adult in the culture of fifty years ago she would have been a daily Mass-goer like her mother. The reason she doesn't go to church now is more to do with contemporary cultural expectation than with the nature of her own belief or disbelief. Nicola is quite clearly a believer. But the pervasive secularity of contemporary life offers little physical or emotional space for the practice of her religion.

Matthew comes next in my rather arbitrary sequence. Unlike Nicola, Matthew is in a strained and uncomfortable dialogue with the church. He knows what he *doesn't* believe in and he sees a lot of that in the spirituality purveyed by the church. He is annoyed with it, because it fails to live up to his expectations. For Matthew, life is complex, it is a struggle and so far he hasn't found any help or any encouragement from the rather superficial versions of spirituality he has encountered in the church. In a way he is disappointed by the lack of religious seriousness in the church. He is asking for more religion from the institution, not less. He also wants a church that is sufficiently open so that he, as an uncertain, struggling searcher, can belong to it without perjuring himself.

Tom is outside the whole Christian ambit. The institution is a very distant presence, hardly even a backcloth. He uses the traditional resources of 'folk religion', sensed as an underlying primeval presence in the culture. Whether his ideas are historically correct or not, he is drawn to the notion of timeless truths that predate newcomers like Christianity. It is not about beliefs and doctrines for Tom; it is about experiencing phenomena that reveal the strangeness of the world. At the same time he is aware of being at odds with secular conventions. Whilst in the right company he is voluble about his beliefs, he is also aware that he needs to be circumspect, especially when his experience conflicts with what he sees as scientific respectability.

Of the four people, Stephanie's spirituality is by far the most purposeful. It is the oxygen, the life-blood of her existence. Although she has chosen to explore spirituality, it is not a commodity for her; it is life. In that respect she is very different from Nicola, who does seem to see religion as a useful commodity or, as she put it, 'money in the bank'. Stephanie has an intense religiosity not unfamiliar within Christian communities. Yet although many of her beliefs parallel those of Christianity she feels no need to belong to the Christian institution, indeed disparages it for what she sees as its false understanding of suffering. Something important is being said here about freedom, about the perceived constrictions of formal religion, seen as an externally imposed set of doctrines and rules to be accepted lock, stock and barrel.

Chapter 4

SHARED ASPECTS OF THE SPIRITUAL QUEST

We come to God in one another's company.

Anonymous Cistercian monk

In the previous chapter I introduced four very different kinds of people. I wanted to show how the uniqueness of someone's life story shapes both the direction of their personal spiritual search and the distinctive way in which they express their experience. Yet alongside the individual differences that make each of us unlike anyone else are social influences that are common to a particular community, felt to a greater or lesser extent by all of those living within it. I emphasise the word 'social' to underline the fact that we are still investigating difference, though now at the communal level. In Chapter 6 I shall move on to the universal, tracking down the underpinning of the spiritual life that we all share as members of the species *Homo sapiens.*

A careful reading through of the transcripts of the research conversations reveals two major social forces common to all the people with whom we spoke. On the one hand there is what I shall call 'cultural inertia', and on the other hand lies the dominance of secularism as an ideology. It is to these general effects of our Western history as they affect the expression of spirituality that I now turn. In the second half of the chapter I will discuss another shared characteristic of spiritual experience, the astonishing emotional power attributed to it by our interviewees and how they responded.

CULTURAL INERTIA

Cultural inertia pervades our lives in myriad ways. It is said that polite forms like shaking hands or raising one's hat have their origins as medieval demonstrations of peaceful intent ('I have no weapon in my hand' or 'I raise the visor of my helmet so that you can see I am a friend'). More often than not, what comes to my mind when I think of inertia is not behaviour but language. Although language is evolving all the time, alongside the changes, and lingering in everyday chatter, are words first used thousands of years ago in languages nobody speaks any more. Religious language is no exception, and the historical dominance of religion is hinted at by the way religious words and phrases continue to be used as a means of emphasis or for expressing shock. People who have no religious belief, or even despise it, continue to say unthinkingly, 'Thank God, I managed to catch my plane', or 'Good Lord, you don't expect me to believe that.' During my student days a famously eccentric university professor had the habit of reading a newspaper as he was walking along the street, inevitably bumping into people from time to time. On one occasion he crashed into a passer-by particularly violently, leading that individual to gasp, 'Jesus Christ!' The professor's riposte, 'Yes, travelling incognito', was met with a stare of blank incomprehension, for the injured man was too detached from the religious context of his outburst to make sense of the joke.

Here is another illustration of what I am trying to convey. Not long ago I overheard some Muslim friends use the phrase 'clapped out' to refer to the Christian religion. Looking at the plummeting numbers of people who go to church it is hard to disagree, but they had something other than attendance statistics in mind. They were struck by the lack of any kind of religious reference in the everyday conversation of most of their non-Muslim English colleagues, which in itself contributed to an uncomfortable cultural gap. At the overt level I am sure their perception is often correct, but let me imagine the following (fictitious) scenario. The reflection emerged one day when I happened to be studying a street map to find the location of the East London mosque and noticed that one of the roads leading towards it is called Christian Street.

Suppose a young Asian Muslim has arrived for the first time in London, where he hopes that an official at the East London mosque has arranged to rent a room for him in Whitechapel. The two are due to meet at the airport, but somehow they miss each other. Having decided to try to make his own way to his lodgings, the young man ends up in Blackfriars tube station late on a Sunday afternoon, confused and hungry. Most shops and businesses are shut, but he sees Bert, an elderly Cockney, standing at the entrance to the station, and asks for advice on buying provisions and how to get to St George's Estate, which he has been told is near his rented room. Seeing how anxious he is, Bert who is a friendly, extroverted sort of man reassures him:

> There's no need to worry, I'll take you there. Cor blimey, I'm not one to pass by on the other side. But for Pete's sake, you need to buy some food. There's a bloody great superstore we pass on your way, down Thomas More Street. It will shut soon, but seek and you'll find, as they say, so gird your loins and touch wood that I'll get you there by the eleventh hour – then I'm off for a glass of spirits up at the White Hart.

I'm assuming what is more than likely to be the case, that Bert knows little of history, is totally indifferent to religion and wouldn't touch the church with a bargepole except possibly to attend his own funeral. But notice the plethora of implicit religious references in the last few lines. Firstly there are the names of places in Bert's home area. As a Cockney, he was by definition born within the sound of the bells of Bow parish church. Whitechapel got its name from St Mary Matfelon church, reputedly painted externally with whitewash when it was built in 1329. Blackfriars takes its origin from the Dominican friary that stood there in medieval times and St George's Estate is named after the patron saint of England. I could have added many other names in Whitechapel: St Katherine's Dock, Hermitage Wall, John Fisher Street, Swedenborg Gardens, Back Church Lane, or, near Blackfriars tube station, streets named after prayers, like Paternoster Row and Ave Maria Lane. The immediate source of Bert's name (even in our multicultural society, still often referred to as his Christian name) is probably the Prince Consort, Queen Victoria's husband, but more remotely he is named after a number of Christian saints, for example, the medieval scholastic St Albert the Great. Secondly, in the space of three sentences Bert makes at least ten references to Christianity. All these are implicitly placing his Muslim companion in the milieu of a complex and many-layered religious history equally as rich as that of Islam – even though he and Bert may both be entirely unaware of it.

If we unpack the content of Bert's words to reach the half-submerged religious references, he is potentially saying something like this:

> May God blind me if I am the sort of person who lets people down, as did the priest and the Levite in the Gospel story who passed by a wounded man on the other side of the road instead of stopping to help him. You need to buy food, but it is evening on Sunday, the Christian day of rest, and it is more difficult to find a shop open at this time. Then for the sake of the good name of the church as represented by the Pope, the successor of St Peter and by the help of Our Lady the mother of God, I can think of a superstore on a street named after St Thomas More, one of the Forty English martyrs. Encouraging myself by quoting Jesus' injunction that those who seek will find, I now urge you to gird your loins, that is, gather your strength together as Jesus advises us in the twelfth chapter of St Luke's Gospel. I metaphorically touch the wood of Christ's cross, asking for divine help so that at worst we will arrive at the last minute. I hope that if we are very late we will nevertheless be treated as well as those labourers in Jesus' parable who only began working at the eleventh hour yet received the same reward as others who had worked all day. Then I'm off for a drink of spirits, representing essence, just as the human spirit represents the essence of who we are. I'll be drinking it at a pub commemorating the occasion when St Hubert (incidentally another possible source of Bert's name), patron of hunters, met our Saviour Jesus Christ in the form of a white hart.

Of course, I have intentionally loaded the monologue to make a point. Bert would almost certainly not recognise himself in this expansion of his remarks and might well feel embarrassed or burst out laughing. Nevertheless, all of the phrases and terms I crammed into his mouth would be perfectly familiar to him. Inertia of language is especially obvious in those quiet backwaters of life that tend to be ignored by contemporary fashion, where many people think and feel in much the same way as their parents and grandparents did. At least in our society, spirituality generally resides in just such a remote corner. As a result, religious words and phrases are still profusely embedded in everyday language, so that even the most secularised individuals pick them up almost without realising it. For most of the time these connotations are hidden or dead. Their significance is that they offer a communal resource to people like those we spoke with during our research who are struggling to express a spirituality that they have almost never given voice to before.

TIMIDITY IN THE FACE OF SECULARISM

Traditional religious language may be a spiritual resource, but getting round to using it is often frustrated by a second highly significant social phenomenon, timidity over admitting that one even has a spiritual life. The dominance of secular assumptions about the nature of reality is a major source of conformity and it pulls very successfully in the opposite direction from the linguistic inertia we have just been discussing. I don't need to consult other people to know that this is so, for my professional role has meant that I am personally aware of the difficulty. In social gatherings I have to brace myself when someone asks me what I do, since I know from experience that the common response when I say that I study spirituality is a pregnant pause followed by 'that must be interesting' and a speedy move on to an easier conversational topic.

The taboo on associating oneself positively with religion is particularly strong in groups of more than two or three people, as became clear to us when we began analysing the differences between what was said in the focus groups and in private. What holds in public for religion usually also holds for spirituality. In spite of the changing understanding of the association between the two, and the positive image of the latter, it is as if most people assume that others continue to recognise the traditional link. The fascinating phenomenon we uncovered was the frequency with which individuals who were searingly hostile to religion in the public forum quite often expressed a much more benign attitude in the privacy of a research conversation.

Twenty out of the 31 people with whom we spoke obviously shared my own uneasiness, for, without being primed, they specifically volunteered the fact that this is a difficult and embarrassing area. The discomfiture has created a conspiracy of silence somewhat analogous to the unnatural hush round sexual matters in Victorian times. In turn, the silence has led to a radical underestimation of how pervasive and important spirituality is to most people. Once the research got under way, two pieces of data made this obvious. Firstly, it was impossible to ignore the finding of our national survey that the great majority of the adult population of Britain *know* there is a spiritual dimension to their lives. Secondly, whatever else it may be, spirituality is supremely related to ultimate meaning. This was a concern of everybody without exception, even those who were influenced by the positivism of the 1950s into dismissing such reflections as useless. One woman spoke for many when she said, 'I think about it [meaning] all the time.' In the permissive atmosphere of a private research conversation, once she got over her initial timidity, she explained how novel it was for her to be able to share her concerns so freely. She added that she would feel completely unable to speak about it to anyone else.

The arrangement we made to meet each person in private and with a guarantee of anonymity was the key to unlocking some very tightly closed doors. Remarkably, in view of our own prior scepticism about people's willingness to share this part of their lives, with the exception of one person every member of the focus groups found themselves able to communicate something of their spiritual life and longings. Paul was probably the shyest of our interviewees. When he talked about formal religion during the research discussion at his home, he began in a defensive, distanced kind of way. He affected a cynical stance, yet as the conversation progressed he revealed that he had a much warmer attitude to the broader question of spirituality, associated with his longing to know. At one deeply moving moment he became filled with emotion as he said wistfully, 'It would be fantastic to know, but we never will.' Then, in line with the second part of his remark, as the time set aside for the conversation neared its close he began to withdraw once more into a dismissive mode, which continued until the meeting came to an end. I found myself wondering if Paul felt he had gone beyond what is socially acceptable and needed to repudiate his concern.

The shape of Paul's conversation was somewhat unusual. In most cases, people's need to protect a delicate area of their lives meant that the longer the conversation went on the more they felt able to be candid. As far as I can judge, the point when it felt safe to talk was when the person was sufficiently convinced that we were neither intent on evangelising them nor secretly criticising them. Once trust had been gained in this way and they sensed that the conversation was drawing to a close, people would often express a curiosity about our own beliefs and experience. In such cases our practice was to give a brief but honest personal statement This usually triggered a sense of relief in the interviewee, perhaps accompanied by further intimate accounts of their experience. Some feel for the delicacy of the subject matter will be evident in the following examples.

When Sarah eventually started to talk about her spirituality to my colleague Kate Hunt, she did so with considerable apprehension. After a long interval and very shyly she told Kate about being aware of the presence of her dead father whilst she was giving birth to her daughter:

> ... when I gave birth to Carrie, I mean I don't know if people will laugh at me, but I actually, I mean I always thought that my father looked after me ... I could see this ... this sounds really funny, but I could see this bright light, and it was like, it was like water and light ... I thought you know, perhaps I'm more spiritual than I actually thought (she laughs). You don't think I'm crazy?

Was Kate inwardly laughing at her? Definitely not, but the fear of such treatment is widespread. Jenny, discussing the focus group of which she was a member, 'thought it was fascinating. I mean religion is such an interesting subject anyway' but 'people tend to sort of shy away from talking about it'. Others found the subject distinctly odd and off-putting. Lucy, whose childhood experience of fundamentalist religion had caused her much unhappiness, chose a colleague who was 'strange' as an example of someone who was spiritual:

> ... he's a really lonely guy ... he's really really nice, but he's just like, he's not in this world almost. And he doesn't know how to react to people. He doesn't know how to have a normal conversation. And he doesn't read any other books apart from his (Buddhist) teaching, and he doesn't watch television, he doesn't read a newspaper. I don't know, I can't imagine a family man doing that.

When asked about a sense of presence, Lucy thought the idea was 'scary ... I think that opens a whole can of worms really.' Debbie, reflecting on the suggestion that some people feel they have sensed the presence of God, said, 'To me, that's going to the extreme.' Tracy's comment about the realm of the spirit was that 'we never really talk about it, you see. We always make fun of it.' Then paradoxically, a short time later Tracy's husband Jack (who had asked to be present during the conversation) said:

> I mean they even killed him didn't they, tortured him. It's a great subject though, religion, isn't it really? You can talk about it all day and all night. It's like politics as well, isn't it, it's a lovely subject ...

A very few people, particularly amongst the over-forties, felt sufficiently secure not to worry overmuch about criticism. And one younger person accepted that her experience of reality was socially unacceptable but she was content to bear the implied sneers of others. I introduced Stephanie in Chapter 3. Her enthusiasm for Universal Consciousness had earned her some odd looks in her focus group and you will recall that she made a remark about 'not being what they expected, a bit of a loose cannon', and added,

> ... but that was OK. I mean, you know, they laughed at people who invented the aeroplanes when they first started, so I don't worry about that.

Tom, whom we have also met previously, was happy to talk about his role as a kind of *Feng Shui* man for his friends when they were thinking of buying a house. He was amused by their embarrassment. They brought up the subject in an extremely oblique way with him, asking if he would mind 'giving the house the once-over', without mentioning things like 'vibes' or atmospheres.

For some people the pains of hiding away or suppressing their spiritual concerns were very real. Joanne, who said that she found any kind of religion interesting, had tried suggesting to her children that they might go to church with her at Christmas, only to be met with ridicule:

> They were horrified. Thought Mum was gone nutty, 'Oh God, Mum's gone religious!' They wouldn't come with me, and they were embarrassed actually. And I didn't go in the end, but, yeh, they were embarrassed ... But it's got a stigma attached to it now, hasn't it, to say religious. Almost as if you're a bit quirky.

Evelyn had a deeply moving sense of the spiritual and was especially stirred by religious art and architecture, about which she knew a lot. She had been greatly affected some years previously when she had got as far as being confirmed in the Church of England, but had not shared her experience with anyone else, nor had she felt able to continue to go to church. Her fear of the jarring contempt of her colleagues is clear in the following passage:

> I don't know whether it's because people think it's a bit phoney or a bit strange, you know. They might say something like, well, 'That's an odd experience. Anybody else got any funny experiences? She's seen the light, haven't you, Evelyn?' And I never really told anybody how amazing it was. Perhaps it's just not trendy. Perhaps people think we're too enlightened now.

Even with near relatives, people tend to keep their thoughts about spirituality to themselves. Graham explained that he had not really got beyond the briefest mention of his spiritual experience to the person closest to him,

> I've probably mentioned it the once to the wife. I think she was a little bemused. She sort of 'Ooh, ooh dear', you know; that sort of thing.

Occasionally there were hints of deeper suppression and here the gender difference is significant. You will have noticed that most of the quotations above come from women. Men in particular had difficulty in talking about or even admitting to their spirituality, though there were plenty of hints that it was there. Simon found the focus group distinctly uncomfortable. He said that some of the conversation was 'way off beat from anything I ever come across or speak to.' He interpreted a question about personal experience as an invitation to 'own up' and said significantly, 'No, I'm not confessing to anything like that' whilst other parts of the conversation showed that he was perfectly well aware of the spiritual dimension. Matthew, whilst claiming to have no religion, described his spiritual life in a way that suggested that it was at least as vivid as that of many church-goers. But he added, 'These guys I play football with on Sunday, you know, they'd think I was a lunatic if they could hear me sort of talking [like this].' Graham had been a professional footballer and recalled how it was standard practice for the lads to talk about religion in a derogatory fashion. Robert avoided talking about the subject at home, but might bring it up with his friends, 'normally when drunk'. For Alan, this was not a subject he could ever remember discussing with his mates, except on one occasion when 'We had a very stupid [argument?] you know, one of those sorts of things you get on to … at twelve-o-clock at night.'

MORE ON THE PERCEIVED SOURCES OF TIMIDITY

In a later chapter I will deal with the historical reasons for timidity about spirituality. Here I want to continue looking at the contemporary context and the rationale for the taboo. One or two people offered an explicit diagnosis of the embarrassment and suppression.

LACK OF SOCIAL PERMISSION

Belinda suggested it had something to do with an unspoken interdict applied to the whole of the secular world. As she saw things, the only people who would feel it appropriate to talk about the spiritual life were churchgoers.

> I mean that's personal. In the sort of circle that I move in and family and what have you (she felt her husband was probably an atheist) I don't mix with a lot of people who go to church.

Stephanie was quite clear that entry into the dominant culture directs people's attention away from spiritual awareness, ' … you are taught things at home and at school which take you away from all that, which is a sad thing.' She also saw certain kinds of language use as creating barriers:

> Yeh, and they tend to put [those kinds of] books, it's sad, in the Library under 'Supernatural'. And well, if you take the two words 'super' and 'natural' separately, they're inoffensive, but put them together and people think 'weird', 'occult', 'strange', which is a shame because it's not, it's supernatural.

Mary, who is Irish and in her sixties, enjoyed *Songs of Praise* on television,

> And that deep feeling of singing to God, you know, and it being all right, without someone saying this is a crank, a religious idiot or something like that. In this country you're a little out of

> step if you go to church. In Ireland you're out of step if you don't.[1]

James, who had by no means repudiated his spirituality, had thought deeply about the role of his upbringing in giving him permission to be spiritually articulate and perhaps spiritually aware. By implication, he sees the problem as being rooted in the lack of an available language, for he interprets what he calls his own 'indoctrination' as freeing him rather than constricting him:

> ... my understanding of another sort of dimension, some sort of spirituality came ... from the beliefs I was given as a child. If I hadn't had those ... that sort of indoctrination, where would I have got that idea from? I mean I can't imagine it just arising from nowhere ... just the same as if I was brought up on a desert island with no other human beings, I wouldn't know how to speak.

FEAR OF STIGMATISATION

Another reason for shyness might be that the spiritual life is simply too personal. In the mainstream of most religious cultures there is a strong tradition of reticence in talking publicly about one's spiritual experience. Even without the scorn of a secularist critique, it involves exposing one's vulnerability. It is not unlike sharing the tender feelings one has for a lover. To speak of such private matters to another person is to give oneself as a hostage to their sensitivity or, more worryingly, their lack of it. It is recognised in evangelical sects, where the giving of public testimony to one's salvation is a cultural expectation, that there is a difficult hurdle to be cleared before it becomes possible. It is here that stereotyped or formulaic presentations can conceal (and therefore protect) as much as they reveal of personal experience. Then again, there is the problem of putting into words what is frequently experienced as being beyond words. Even those who have been most garrulous about their spiritual life, like St Teresa of Avila, reach a point where they are lost for words.

These are understandable reasons for reticence. But they are not the best explanations for the shyness of the people with whom we spoke. In 1985, along with my colleague Ann Morisy, I published the results of an in-depth survey that revealed by far the most important reason for timidity. We were making a study of the spiritual life of a random sample of the citizens of Nottingham. Amongst the experiences our interviewees spoke of were being aware of the presence of God, of discovering the ultimately real, of discovering that one is an intimate part of a larger reality, of finding meaning, of being comforted in sorrow, of being given strength or delight and of being given a moral call. In essence, they claimed to come to know by acquaintance something about reality that they did not know before. Almost all of them were positive and appreciative about this dimension of experience. Some said it was of pivotal importance to their lives. We were therefore very struck by the fact that as many as 40 per cent of our informants in the 1985 study said that before confiding in us, they had never told anyone at all about their experience; not even someone as intimate as another family member or partner. Even those who *had* spoken admitted that they had been pretty shamefaced about it, rather like Graham whom I mentioned a moment ago. When we asked about their reluctance, everyone without exception said they feared being labelled either stupid or mad.

We made another curious discovery during the 1985 research. We had noticed during pilot research that occasionally someone would say they thought people who denied having had any spiritual experience were lying. So in the main research we decided to probe this with the following question:

> What sort of person do you think of as claiming not to have ever had such experience?

We had expected a range of opinion from 'rational', intelligent, sensible, etc., through to a few more negative evaluations. In fact the judgements were startlingly and overwhelmingly negative, to the extent that we compiled a list that I still have in my possession. According to our informants, those who claimed they had never had such experience were,

> Apathetic, bitter, conformists, cowards, dull, emotionless, hard, ignorant, insecure, insensitive, know-alls, lacking capacity, liars, materialists, mean, miserable, morally lax, narrow minded, over-controlling, sceptics, self-centred, sneerers, superficial, too busy, unaware, unimaginative, unintelligent, unpleasant and weak!

In defence of those people who deny having had any spiritual experience, I need to add that I have personally never found any direct evidence whatsoever that members of this group resemble this spectacularly unpleasant stereotype! The typecasting on the other hand suggests a great deal of anger on the part of those who claim a spiritual dimension to their experience. It seems that their frustration is projected onto 'them', the anonymous mass of the surrounding society whom they see as refusing them permission to take their perception of reality seriously.

That is not to deny that there are highly influential authority figures in contemporary society who give good grounds for people to be apprehensive about how their spiritual experience will be received. The feeling that it is 'not allowed' gets confirmation in the psychiatric profession, where to this day there is a good deal of ambivalence about whether or not such experience is evidence of mental disorder. Among the eight criteria in the *Diagnostic and Statistical Manual of Mental Disorders* of the American Psychiatric Association (DSM),[2] any two of which are considered to be enough to identify the possible onset of a schizophrenic disorder, are the following:

1. Odd or bizarre ideation, or magical thinking, e.g. superstitiousness, clairvoyance, telepathy, 'sixth sense', 'others can feel my feelings', overvalued ideas, *ideas of reference* [for example feeling, like the poet Wordsworth above Tintern Abbey, that there is a presence 'rolling through all things'].
2. Unusual perceptual experiences e.g. recurrent illusions, *sensing the presence of a force or person not actually present.*

In the diagnosis of Schizotypal Personality Disorder, the same criteria are used, with an example of what is meant by sensing the presence of a person not actually there, e.g. 'I felt as if my dead mother were in the room with me.' In another form of psychosis, manic disorder, 'God's voice may be heard explaining that the individual has a special mission.'

These criteria would be satisfied by very many of the accounts of religious experience in the archive set up by Alister Hardy. It is true of course that the way a person suspected of being mentally ill presents themselves is also an important criterion, but there is still an uneasy boundary here. Not so long ago in the 1987 edition (now superseded) of DSM, there was a passage where the psychiatrist was given the following advice on discriminating between true psychosis and culturally mediated beliefs:

> Beliefs or experiences of members of religious or other subcultural groups may be difficult to distinguish from delusions or hallucinations. When such experiences are shared and accepted by a subcultural group they should not be considered evidence of psychosis.[3]

That is to say, certain kinds of experience are not to be taken as evidence of mental illness when they occur in a cultural context that expects and gives permission for them. But those same phenomena become possible criteria for the diagnosis of mental illness when they are manifested in a group that does *not* give social permission for them. Thus, if a native American talks about meeting a spiritual presence during a tribal ritual, that is normal. It is part of the standard belief system of the tribe. On the other hand, if a New York stockbroker says that he has been aware of a spiritual presence, then the psychiatrist needs to be suspicious. Why should this be so, if such awareness is natural to the human species? It is significant that this particular note has been dropped in subsequent editions of DSM, but clearly the issue is still an uneasy one. Are the categories of sanity and insanity adequately determined by what is socially permissible in a culture? If so, this is reminiscent of the Stalinist practice in the Soviet Union of defining people who disagreed with the political tenets of Marxism/Leninism as mentally ill.

THE POWER OF SPIRITUALITY

The sheer bodily animation and the energy in people's voices when they talked of their spiritual awareness made a strong impression on Kate and me. As we sat and listened, it often felt as if an overwhelmingly powerful reality was lying like a mighty headland behind a mist of timidity and denial. Another metaphor that comes to mind is Gulliver's imprisonment in Lilliput, tied down by a multitude of slender cords. The threads are made up of repeated scornful dismissals in the media and in ordinary conversation, plus a host of vaguely sensed doubts about the plausibility of religion, assumed to be the normal vehicle of spirituality.

Quite often the threads of scorn have been perversely created by the triviality or restrictiveness of a spiritual tradition itself, as purveyed by a school or religious institution during childhood. Lucy, who felt her life had been blighted during her teenage years by enforced attendance at a fundamentalist church, remarked that she 'would like to have something a bit bigger ... a good [deal] more real'. In spite of the rejection of such received ideas, spiritual matters continued to preoccupy her. Jenny, who said she was 'certain in my own mind that there is no God', nevertheless added, 'It's something you think about on a continual basis anyway.' Sometimes there was a sense of being driven, even against one's better judgement. Paul was sceptical of ever reaching the nature of ultimate reality,

> Well yes, [it's] one of those great questions that everybody asks. I mean if you had all the answers, what knowledge [that would be], absolutely fantastic knowledge ... was there a God, you know what I mean. But we'll never ever find out, I'm sure we won't.

His painful longing for an insight into the depths of existence led him to search far beyond the religious institutions, alert for the slightest hint coming from any direction.

For other people, in spite of their scepticism, this vehement interest had been followed by a direct awareness of the spiritual dimension. In a sense, they had sought and they had found. What, in the opinion of our informants, precipitated the shift? The consensus was that there were two major influences – the violence with which extreme situations break in on one's sceptical presuppositions – and the seductive beauty of the world, even in the midst of distress.

MATTERS OF LIFE AND DEATH

We have seen how Stephanie linked her spirituality to a consciousness of feeling somehow more than her body, which she was very aware would one day die. But as she said, 'you are taught things at home and at school which take you away from that, which is a sad thing.' In Stephanie's case, the 'taking away' did not in the end succeed, because of the impact of the brute facts of her life:

> I've had some quite devastating romantic occasions. I used to have this 'bastard magnet', if you'll excuse the phraseology and I always used to find the men who would hurt me ... I mean

> I'm not talking about got stood up, or anything like that ... And I think a lot of people turn to, like, religion or spiritual growth because of pain, is basically what I'm saying.

The implication is that the spiritual world may be covered over by the assumptions of secular education but it continues to be there, just under the surface. Debbie was representative of most people when she talked about spirituality as 'something that comes from the depth within'. She worked with cancer patients and was greatly moved by the suffering she encountered:

> I think it's the mental anguish that a lot of them are going through as well as the physical pain. A lot of them are very frightened people. I mean I particularly recall one man who was terrified, he just didn't want his wife to go out of the room, didn't want her to go down to the shop, didn't want her to go anywhere. And because I was there ... I think he looked upon me as well, at least Debbie's here.

Debbie was in her early forties and felt it was her religious upbringing that enabled her to do her sometimes harrowing job. It was here that she turned when she wanted to give and receive strength. The crucial nature of her religion in sustaining her in her career nevertheless did not express itself in attendance at a place of worship.

When there is a health crisis, spiritual support can sometimes seem more important than medical support. James was in no doubt that the loving presence he encounters in meditation deals with fear better than the medical profession:

> Well, certainly when I thought I'd had a heart attack and I was in hospital in Accident and Emergency ... then I was very much thinking 'I need you now', and that was quite vivid. Although [I have] a sense of embarrassment now to recognise that, almost. I think it was about 'You can save my life now. You can keep me alive. You know all this paraphernalia around me is scary and I'm looking to you to help me, not to the doctors in their white coats.'

James was well aware of the reductionist interpretation that explains away such experience as an illusion brought on by fear. To a degree he subscribed to that view as it applies to conventional prayer, but then went beyond it:

> Prayers and things like that seem empty and futile, but I have a sort of a more direct relationship with something intangible … It's a feeling, something intuitive, it's something that's so personal. To me, it's beyond question.

Carol was openly fearful of death: 'What frightens me is the fact that if your mind's still there [after your body dies].' For her, death was equated with being cut off from relationship, and she remembered her repeated experience of being amongst crowds of people in the middle of Nottingham:

> I used to go into town and just look at the whole of town all full, and just think, in a hundred years all of these people will be dead …

In contrast with the fear of being cut off in death, spiritual awareness was often described as transcending every kind of isolation. Stephanie meditated regularly and spoke with deep emotion about its relation to suffering. She spoke of responding to her agony with 'deep wailing sobs' but in her opinion many people who are in pain block them out because 'that's what society says we should do'. For Stephanie, to permit herself to experience the pain of life, in the context of meditation, is also to 'allow the universe in'. It is a two-way process:

> I mean, you should feel it sometimes, because I believe that with pain there is no way round it. You can't go over it, you can't go under it. … there isn't a side door, you've got to go right the way through it. So if I'm on my own, I let it out, so that I can be filled back up with the positive things, with the universal consciousness, to be in tune with everyone.

In the case of older people, it was as if advancing years brought on a return of suppressed (or perhaps repressed) questions. Joan talked about the effect of retirement:

> For the first time in forty years there's just the two of us … and you do have more time to talk to people and to see people and to think … there's a lot of people of our age group that that

> happens to, and of course you tend to think of your own mortality as well ... their parents have died [and other relatives] so you're coming in contact with it a lot more, so it gives you the chance to talk about it more.

THE SEDUCTIVE BEAUTY OF THE WORLD

These confrontations with one's mortality, or the experience of being profoundly alone, may strike us as encounters with the existential *angst* that the philosopher Martin Heidegger recognises as jolting people out of inauthenticity. But there is another kind of experience at the other end of the emotional scale which also wakens awareness. Alongside the many distresses in her life, Stephanie talked about the beauty and spiritual seductiveness of the world:

> The universe is a wonderful, bountiful place and it will give you whatever you want, provided you send out a clear enough message.

This sense of connection with that universe, of experiencing some underlying coherence, appeared in Debbie's account of a world filled with meaning, even to the extent that a leaf falling off a tree has significance. She described a visit to the grave of a friend (Jack) who had died young:

> I was sort of speaking to him with my husband ... around his grave and I was just aware that something had happened, some leaf had come off a tree just in answer to something I had said in jest ... and Jack had a great sense of humour ... it was just an acknowledgement, I felt.

Later when she was walking in the hills and talking about Jack, the same thing happened, 'A leaf trickled down in front of me and fell to the floor ... even my husband picked up on that.' Such occasions of meaning-filled presence are often filled with joy, to the extent that the person is moved to tears. Thus another of our interviewees, Joan, recalled an occasion in a church many years ago, when she very clearly sensed the nearness of God and was 'feeling very tearful'. Her tears were a mixture of delight and distress, because at the time she was worried because her husband was in hospital following an accident.

Matthew was swept away by the splendour of the human story. He had a heroic vision of history that made him feel that at some level there has to be 'more':

> I love words, and literature, and philosophy, and the world religions and the demonstrations of human intelligence in all its manifestations ... I don't think we erect these splendid edifices, well we do, but I don't think we [ourselves can take the credit?] ... and it's just layer upon layer, generation upon generation, each one that washes forward another, you know, recedes back into the sea, and there's some more important shells and pebbles on the beach.

Even in the act of disparaging what he saw as the petty restrictions of the Christian orthodoxy to which he had been introduced in childhood, Paul implied an experience of all-embracing coherence:

> God's everywhere. God's in that tree, God's in that fish, God's in that whatever. So why not pray to that tree? Why not sit next to a tree, and why go to a church? ... I could make a cross out of two twigs ... it's only a symbol.

To summarise, in spite of being disappointed by what purported experts have to say of the realm of the spirit, the direct intuitive encounter does not disappoint. James gave up religion after being disillusioned during the study of theology at university, but,

> the God bit didn't leave me. So there was this sense of ... a Being, something sort of ethereal, superior almost in a ... pure way. And that's still around me now.

RESPONDING

PRAYER

Our informants were consistent in telling us that distressing experience of the kind I illustrated a little earlier almost always elicits an involuntary crying out, whether silent or out loud. At such times even self-proclaimed atheists sometimes find themselves, for a brief unguarded moment, engaged in prayer. The major division between people is not whether they do or do not cry for help *in extremis*, but how they interpret what they are doing. Probably the best-known sceptical explanation for religious adherence is the 'deficit theory'.

This explains people's propensity to engage in religious behaviour like prayer or ritual as due to some lack of comfort or deficiency in life such as loneliness, illness, physical danger, great pain, or the fear of growing old and incapable. Arthur Hugh Clough's verse sums it up:

> And almost everyone when age,
> Disease, or sorrows strike him
> Inclines to think there is a God,
> Or something very like him.[4]

Several of our sceptical interviewees fitted this pattern, turning to God in a crisis, even though at the rational level they were quite clear that what they were doing was unreasonable. Lucy interpreted her prayer as a conditioned reflex, a kind of psychological tic left over from childhood teaching:

> … I don't believe, well I can't even say that I don't believe in God, because I don't know if I do or I don't, because it's rather difficult to let go of the religion that you had when you were younger.

She saw prayer as a weakness, one that she could rather easily fall into, as may have happened when her grandfather was in hospital:

> I can't remember praying for him, but I wouldn't put it past myself to have done it – put it that way.

Matthew, though he was almost as sceptical as Lucy, found a degree of comfort in putting his anxieties into words:

> When I feel probably not very happy, and you just have a talk to yourself, and have an inner voice chattering away to you … kind of puts things in order and, again, probably some kind of meaning in things … and sort of wishing for happiness … hate to say 'asking for it' because then we're starting on praying, but yes, sort of offering something up to something that probably isn't there.

Trying to remember occasions when she had resorted to prayer, Joanne recalled how she had narrowly missed being involved in a serious crash when she was driving, with her children in the back of the car. She found herself spontaneously offering a prayer of thanks to God, though with an awareness that she was probably being superstitious:

> … one of the kids was crying and I turned round to look at [him?] as I was going under this bridge, and something just made me turn back, and somebody had broken down in front, and I would have gone into the back of them. I mean you are talking split seconds. And afterwards I really was absolutely, you know, I had to pull up, because I was just so shocked. I thought, 'Thank you God, thank you so much.' Yeh, it's things like that where I've thought, 'Yes, something's watching over me, something won't let me die' … there's a big part of me that feels I'm not meant to go yet.

For other people prayer was a more frequent and authentic part of life. Emma, who was in her twenties, had broken away from her strict Baptist upbringing. Nevertheless she had a strong and quite conventional prayer life. As a child she was taught to say prayers at bedtime, but her prayer nowadays took the form of conversation at any time of the day:

> I mean quite often it will be in bed, but that's purely because that's the only time I've got to … think about the day's events … I'm grateful for everything I've got and I do say that … But you know, if there's things troubling me or whatever then, yes, I'll talk about them as such.

In Mary's case, prayer might be said to be the atmosphere of the whole of her life, in spite of the fact that she did not think of herself as religious:

> Yes, God's important to me; I talk to him every day. I go out early in the morning to work and I'm walking down the street and before I get to the bus stop, I say dear God, thanks for a nice day. It might be pouring down with rain at the time … it's like an invisible somebody walking beside me. So I don't think of myself as being religious but I think of myself as always having him there.

Her wholehearted spirituality did not prevent Mary's attitude to the religious institution from being highly ambivalent. She felt she was 'brainwashed with religion' as a child, yet her prayer had carried her through trauma caused by a member of the family who was undergoing terrifying mental breakdowns:

> I was in a terrible state of distress and I used to feel … I would get from one place to another and I wouldn't know how I got there. I felt I was losing my mind. I was losing all control and I was so desperately in need of help but couldn't talk to anyone. And I used to wake up every morning and I used to say 'Please God, help me just for today.' And that's all I asked and he did … I mean if I didn't have God then I know I wouldn't have got through that …

The legacy of Mary's experience of help during this nightmarish time of her life was her ongoing and profound trust in God:

> I still totally believe that he's watching [over me] all the time and I believe that if ever I'm in desperate need of him he'll be there to help me … I still have daily contact with him.

Petitionary prayer, or even the prayer of thanksgiving, is not the same thing as meditation or contemplative prayer, though in people like James and Mary they tended to merge into each other. In a sense meditation, seen as allowing oneself to be immersed in the immediacy and totality and mystery of one's existence, is potentially a part of everyone's life, and these two people had made it central to their everyday practice.

We have seen that Stephanie had adopted a particular devotional practice as a consciously chosen technique. She advocated it to others as a way that leads towards wholeness and personal healing. Whether this could be classed as prayer or meditation is not entirely clear. In the course of her personal conversation with my colleague Kate she talked extensively about her practice of 'letting the universe take care' of life. Kate commented that from her (conventionally religious) background that sounded very much like praying to God. Stephanie was quick to suggest that this was no more than a problem of semantics:

> Yes, but you see God is just a word at the end of the day, and we might be believing in exactly the same thing, but like I say, we've got different ways of worshipping and a different way of

> getting to that same place … One day we'll both get there and go, 'Oh, hi, did you enjoy the trip?'

Tom was in some ways reminiscent of Stephanie. Though he was quite clear that he was an atheist, he nevertheless spoke of praying when he was in difficulty. To whom or what was he praying, then? The way he described his prayer suggested that it had a strongly contemplative quality. He believed that this act put him in contact with the cosmos of which he is a part and helped to realign him with the universal harmony.

Reflective awareness of the here-and-now was an evident presence in the lives of most of those with whom we spoke, though it did not necessarily have any conscious association with religion. For Joanne during her childhood, the mystery of life opened up for her when she went swimming with her religiously sceptical father on Sundays. As they swam up and down the pool, or splashed about in the shallow end, he would say things like, 'It's all a load of rubbish, all these churches being built', but he also,

> … talked about space, and eternity and things like that and he used to really get me thinking, to a stage where I couldn't conceive eternity or something going on for ever.

Jenny was certain in her own mind that there is no God. But she was concerned about relationship in depth, which she saw as necessary for human survival:

> On the surface we might have those human connections, but without that deeper connection with somebody it would be very shallow … life would be quite meaningless if we didn't have that with people … also it's how you feel within yourself as well, it's not just that connection with other people … this is why I do actually sit and I do spend quite a lot of time on my own now … I find that vitally important, sort of general well-being and happiness.

CONCLUSION

Our informants were caught between making use of the vast legacy of religious language, much of it feeling like wandering through a beautiful graveyard, and sticking to the businesslike sceptical practicality of daily life. At some level everyone we spoke with continued to try to make coherent sense of their experience of life. The longing for insight was inescapable, hammering insistently at the awareness, whether or not they believed in such a thing as ultimate meaning. Intuitions of a spiritual dimension to reality were particularly likely to burst out in two kinds of extreme situation, deep distress and overwhelming delight. It is as though the constraints of a lifetime's conditioning fall away in the face of unavoidable immediacy and, like it or not, the immediate response often amounted to prayer. None of our informants disputed the power of those moments, but when they came back down to earth and were once more enveloped in their everyday cultural matrix, the questions gnawed away: What was that? Something? Or nothing? In the next chapter we shall see what they thought.

Chapter 5

THEORISING ABOUT THE SPIRIT

God guard me from those thoughts men think
In the mind alone;
He that sings a lasting song
Thinks in a marrow-bone.

W. B. Yeats[1]

If you meet the Buddha on the road, kill him!

Zen saying

Ah, love let us be true to one another! …

Matthew Arnold[2]

Every time we give tongue to any subject under the sun, our words contain assumptions about the way the world is. In the previous two chapters I introduced a group of our contemporaries talking about their spiritual lives. I also discussed some important social factors affecting what they are prepared to say and how they say it. But how do they make sense of it? What are their theories of spirituality? This doesn't just mean explanations given afterwards, when they have finished their stories. A study of the transcripts of the conversations shows that implicit meanings are already built in to the language they use, even before they start interpreting, because language is always already 'theory-laden'. These ever-present embedded meanings have very longstanding origins in the history of the culture to which we belong. At a more directly personal level, they are also shaped by what we have assimilated from our parents and teachers in childhood and by the unfolding pattern of our lives. Meanings of this type, that we haven't consciously thought about, have great power simply because we are unaware of them. It follows that if Kate and I were to do our job properly we needed to try to overhear not just description, but something of these deep axioms.

Beyond this implicit meaning-making – and strongly influenced by it – there is a more conscious and explicit level of interpretation of our experience, the sorts of ideas we come up with when we stand back and ask ourselves, 'What was that?' We were particularly interested to hear what people had to say at this explicit level, because it is the plane on which we ourselves were trying to operate as researchers attempting to understand. We wanted to know how much conscious theorising our interviewees had engaged in regarding their spiritual lives. Which interpretations struck them as plausible? Where did they finally turn for an explanation? Most people volunteered their opinions on these questions as a matter of course during the research conversations and if they did not, we invited them to do so. I now turn to a review of what they had to say.

RELIGION

In spite of the data I have quoted on the speed of secularisation in Britain (but in tune with my remarks about cultural inertia), Christian interpretations hang on surprisingly tenaciously in all age groups, though as might be expected this was more obvious in the case of older people. A small cluster of those in the latter group spoke in a consciously Christian manner and made statements of belief that were doctrinally no different from those of practising believers. In one or two cases their theological literacy was a good deal more sophisticated than that of the average churchgoer. They were not offended by the word 'religion' and they envisage their spiritual life entirely in those terms. These people also held strongly to the traditional image of religion as having to do with virtue. Thus one elderly woman wished to be thought of as religious specifically on the grounds that she saw herself as a 'nice person'. Her opinion is in tune with remarks that have an old-fashioned flavour in our multicultural society, as when someone says of bad behaviour that it is 'un-Christian'. The only recognisable feature that differentiated this group from the formally religious was their choice to be absent from the pews on a Sunday. Their refusal did not represent a rejection of religion, but in many cases could be traced to a history of painful mistreatment by one or other of the religious institutions.

Compared to the theological sophistication of some of the older people, the younger members of the sample usually had a much more fragmentary grasp of Christian language. Nevertheless, as we shall see, consciously or unconsciously they almost always turned to it when attempting to interpret their spiritual lives. The combination of distancing oneself from the institution on the one hand, and on the other hand reaching towards religious terminology for explanatory purposes, is bound to result in feelings of dissonance. During the research conversations the symptoms were obvious, in the form of disjointed, stammering speech and an embarrassed tone of voice.

When religious language was necessarily but reluctantly turned to, the effect was to distance the person from the full connotation of the words, implying a weakening of traditional meaning. For example, when someone said they believed in God, this usually proved on investigation to be a vague 'generic' God, rather remote from the Trinitarian and personal God of Christianity. References to Jesus as the focus of the spiritual life were more or less limited to people who had had a childhood background in Christianity. But even they were quite often dubious about traditional orthodoxy, as the following quotation from Emma illustrates:

> [talking about believing in Jesus] Yes, yeh, um, I think that's quite a difficult really, yes, he does, yeh, I mean because I believe in the um, you know, in Jesus as such, he came down and, I don't necessarily understand it, but I, you know, believe it to have happened as such. I wouldn't be able to pinpoint a role for him at the moment you know, I don't quite know what he's doing now, what he's got on his c.v. as such, but yes I do.

Emma's confusion and evident discomfiture, covered over by humour, is very characteristic and suggests the strength of the taboo on talking about religion in contemporary Britain. Compared to Emma, many members of this group were at a significantly greater distance from the institution. Even so, the appropriation of half remembered elements of the Christian tradition was still obvious, though buttressed by a mixture of ideas drawn from other sources such as Eastern religion, spiritualism, paganism and science fiction.

The unmistakable esteem for the spiritual life may have motivated a resort to religious language to protect it, but the usefulness of the language seldom implied a benevolent view of the religious institution itself. Our interviewees thus provided rather pure examples of the spirituality/institutional split diagnosed by Zinnbauer and others.[3] In Chapter 10 I will return to a detailed catalogue of the alleged deficiencies of the religious institution. For the time being I want to set aside these criticisms to concentrate on people's search for plausible ways of interpreting the spiritual life.

TRADITIONAL RELIGION AS AN ADEQUATE SOURCE OF MEANING

About half of the group had an interpretation of reality that coincided fairly closely with certain mainstream Christian ideas. Many of these people gave the impression of having passively absorbed those ideas in childhood and subsequently spending little time consciously reflecting on them. Their orthodoxy functioned at the implicit level I referred to at the beginning of the chapter, by providing a ready-made framework of meaning and a sense of identity.

These implied meanings could operate very powerfully, as in the case of Elizabeth who was Scottish. Her father came from a Protestant 'Orange' background and had had to change his religion to marry her Roman Catholic mother. She had in consequence been brought up as a Catholic and this had caused bitterness amongst her father's relatives. She was well aware of the sectarian rivalries associated with being raised in a strongly Catholic environment in the industrial lowlands of Scotland: 'Our non-Catholic cousins went to different schools and there was still that rift, even in the family.' Her theological framework was unsophisticated and completely traditional and she admitted that it represented no more than what she had learned in catechism class as a child. Although nowadays she never went to church, Elizabeth was quite clear about her tribal identity: 'I'll always be a Catholic, yes, whether I went to church or never went to church now.' For the same reason, at the time we spoke with her, her children were attending Catholic schools.

Sean was one of the youngest members of the group. His parents were Irish, and like Elizabeth, he also saw himself as completely a member of the Catholic 'tribe'. His talk was littered with references to priests and nuns who had been involved with his upbringing, and he remarked, 'I have been planted with that seed.' However, he was not much impressed by religious formality. He said that when he went up for Confirmation at the age of sixteen he 'felt like a battery chicken' and that going to Mass 'isn't of interest to me any more'. But for Sean being a Catholic is more than going to church. He spoke of religious issues turning up for him nowadays particularly in drunken discussions with his friends, sometimes lasting throughout the night. Whilst he described himself as leading a disordered life, he spoke of praying often, and believed that God hears all petitions. His personal prayer to God was movingly reminiscent of the words of the publican standing at the back of the synagogue:[4]

> I know you accept me for who I am. I want you to know that I am thinking about my actions, and I don't want to do bad by the next person.

Occasionally people who in other ways were remote from the church made use of highly conservative religious imagery. Most often this was the case when dealing with children's naïve but searching questions about profundities. Steven had been faced with an impossible query from his six-year-old son who 'automatically thinks everyone goes to heaven', but wonders where heaven is:

> Our dad's dog, he's died and I think one of the comments he made when my auntie died was, 'Well, she can see granddad's dog now.' He's asked us a couple of times where heaven is. What do you say, you know? It's up in the sky, it's got to be up in the sky somewhere because he thinks the devil is down below …

Steven's response was in terms of the ancient picture of a three-decker universe. He was perfectly clear about the mythological nature of the imagery, but felt it was appropriate as a means of offering some kind of answer in the presence of the mystery of death.

As the most extreme example of a break in relationship, death and reflections on the briefness of life were particularly likely to bring out traditional beliefs. Certainly many of the group had a belief in life after death. Gary's mother died when he was a small boy, but he was quite clear that they would meet again:

> I truly believe that one day, when obviously my day comes, I shall go to meet her again and she will be as I remember her. Not just my mum, you know, your grandad, your grandmas, your sisters, your brothers. That's how I like to view it anyway. Not a negative thing – once you've been buried, that's it. No I

don't believe that. I think that's a nice thought as well.

Emma struggled with the ambiguity of metaphor. On the one hand she had a very physical and traditional understanding of heaven and hell:

> I see [heaven] as a real place, yes I do. That might be very naïve and very simplistic of me. I see it as a place where whatever you want it to be, whatever your personal circumstances are, you see whoever you want to see. I mean it sounds very fairy tale-ish, doesn't it, but that's how I see it … And I suppose I see hell as like the burning fires … Eternity burning in the fire.

On the other hand when she was asked if she ever imagined God, she said amid laughter,

> I imagine him to be like he's old, this bloke, with a big beard sat on a cloud. In a sense I know that's ridiculous. So if you have to picture him then that would be how I'd picture him. But that's completely bizarre …

Belinda grappled with intangibles and was very obviously concerned with ultimate meaning. She had pondered the unimaginability of nothingness, 'I can't imagine what it would have been like if we hadn't all been created.' She felt certain there was 'something there', 'otherwise we'd all be floating about in the wind.' We are here for a purpose: 'I think everybody does what they do because that was … designated for them.' She did not depend on concrete imagery, and made a distinction between Jesus and God:

> I can relate to the stories of Jesus as probably living on this earth as a human, as we do. But God's a different thing altogether because he's Life. I do believe that there is some force or whatever you want to call it. I believe he's there but I don't know what form he takes.

Evelyn, perhaps because of her love of art, was sensitive to the power of iconography, particularly when it was tied into the culture of her upbringing:

> If I'm in church or I see religious pictures, they do stir something in me … I do feel more religious when [I] see pictures of the crucifixion than I would of a picture, say, of Ganesh [the

> Hindu elephant god]. I see them [the latter] as religious pictures but they don't stir anything in me.

Nevertheless she was perfectly well aware that anthropomorphic images of God wouldn't do, 'an old man sitting in heaven ... I don't really go along with that sort of thing.'

We have seen that Mary took traditional Christian theology very seriously. Her story is particularly interesting because it shows how religious theory, which in the case of Elizabeth operated primarily by providing her with a tribal identity, can become the mediator of immediate spiritual encounter. Mary supposed that 'really as a child I was brainwashed with religion', but added, 'I accept it and I embrace it in that perhaps I need it'. She had been through great distresses in her life and turned constantly to God in prayer: 'I relied totally on God because I couldn't go out and tell anybody about it.' In the process she felt she had found out something about the nature of God. In her experience, 'he gives us enough strength and he doesn't give us too much, he just gives us enough to carry us from day to day.' As a child she had been disturbed by the idea of God as prying into one's private life:

> You couldn't ever get away from him no matter where you got – you got into a cupboard, God was in there. I remember sitting on the toilet once and crying because I thought he was watching me and I didn't want him to watch me and I found it very disturbing. It was that sort of notion of God as a sort of pryer or a peerer at you.

She had discovered in the course of her very active prayer life that whilst God was always with her he was not the voyeur of her childhood fears. One might say that Mary's beliefs had grown and become more deeply part of her through intense reflection on her life's experience in the light of the religious teaching she had received in her youth. Because of her sustained attention to traditional theology she was able to make it spiritually creative for her.

TRADITIONAL RELIGION AS INADEQUATE

The other half of the group we spoke with quite clearly diverged from traditional Christian orthodoxy in their theorising on the 'big questions'. There were many sources of their alternative ideas, motivated predominantly by the need to put something in the place of religious interpretations discarded as implausible or unpleasant. Yet the hovering presence of traditional religious imagery was everywhere apparent. Alan, who felt 'there's got to be something more to it than just this', scornfully rejected literalism, 'an old chap sat there, pearly gates ... a burning pit and all that sort of stuff'. Sometimes, as was the case with Paul, rejection of tradition stemmed largely from disgust with what he had seen of the religious institution. That did not prevent him from speculating on the nature of God, 'I'm sure if there is a God, he's not going to shove everybody in hell, is he? What sort of God would do that?' Seemingly drawing on theosophical ideas, he wondered out loud whether our time here on earth is,

> ... a stepping stone in evolution, so to speak, that you're here for your sixty, seventy years and then you progress to a higher level, a higher plane, or a different dimension.

For Paul, speculation of this kind was more of an intellectual game than an existential necessity. It fell into the same universe of discourse as the question of the existence of aliens or how the Egyptians created the pyramids; that is to say, the world of remote curiosities. On the other hand he was pretty sure that we are governed by fate.

Tom was another believer in fate. His avowed atheism also included the opinion that all things are combined in an invisible unity. It is this unity that explains the power of premonitions and Tarot cards, the phenomena associated with Ouija boards and the exploits of spiritualist mediums. The source of his views lay in what he took to be a hidden, ancient folk-religion handed down through his family for generations. But he also found spiritual nourishment in the films of the *Kung Fu* adept Bruce Lee, who was perhaps one of the sources of Tom's much-admired brother's belief in the concept of the holy warrior. All of these examples of tangible spiritual phenomena were, in Tom's opinion, evidence of the reality of the world of the spirit and were on a par with the empirical evidence demanded by scientists.

The interconnectedness of things was also important to Colin as, once again, was the notion of fate. His cosmology was gleaned primarily from magazines about the 'X-Factor', watching *Star Trek* on television, and reading science fiction. He was impressed by the speculations of physicists on the existence of parallel universes, and he believed in a life force that moves on to another universe when we die. Death is in fact a new beginning. Death can also give us insights that, in this universe, are normally reserved to people with 'heightened senses', by whom Colin means clairvoyants:

> Your body dies and you leave it, I think that's the only time you probably see what's going on. You leave the body that is dead but you're still alive when you leave the body, but you can't interact with the present.

There are no such things as ghosts but there are electrical fields that can trap the life-force. The parallel universes could sometimes interact and sometimes you might get a crossover … That's where you get your ghost images.

His view of the origin of the life-force was:

> I think it's just something that's there. It's just something that's gone on and on. I don't think anybody knows really where it all started. I suppose you could say it's from the Big Bang you know, electrical things and … molecules come together …

There may be a gender bias here, since all six of the people who turned to the language of science for their interpretations were men. Colin's metaphors are very obviously borrowed from science and science fiction and whilst they may strike an outsider as lacking credibility, they utilise the high status of scientific language to gain plausibility. Another example is John, who in many ways clung on to Christian orthodoxy, but had no time for anthropomorphic imagery, preferring to use familiar metaphors drawn from physics. God is:

> … just a force, an energy if you like, a driving force. I do feel that there perhaps is more than that to what God is, but I can't visualise or see what that is.

Other people wanted to get away from imagery altogether. Sharon was one such person. In spite of her disclaimers her concepts seemed to carry a degree of theistic belief:

> If you're a spiritual person, you believe there is something there, controlling us if you like, or watching over us, or dictating the path that we take, not God, not ghosts, or spirits, or anything like that. It's not a religion in the sense that you're actual-

> ly worshipping something or someone, it's the sense that there's something definitely there that we don't understand.

Matthew was a theologian by temperament, although he had never made a formal study of the subject. The closest he could get to saying something specific about God was to suggest that 'he is mind'. His views were universalist:

> We are here ... I believe that all of us should treat everything with tremendous respect ... I think this comes down from my problem [of] embracing 'that bunch of people that worship in that place'. I'd like to hang out with them all.

The struggle for Matthew was with paradox; holding opposites together. In thinking about transcendence he would have preferred to use no imagery, whilst at the same time realising that we cannot do without it. He admired believers who held on to the tension between belief and unbelief. He himself had had virtually no religious upbringing and the sceptical side of him was strengthened when his childhood friend died of leukaemia:

> It confirmed the meaninglessness of things, just reinforced a prejudice I had that this isn't adding up to going any place – it's such arbitrary, chaotic tragedy and nonsense.

In Matthew's opinion, fear creates the gods and yet the splendour of human existence pulls us very strongly into believing in some form of transcendence. It is almost like an animal instinct:

> I think the reasons they erected these splendid edifices [are to do with] a very primitive set of beliefs. But I wasn't talking just about churches and cathedrals. No, I was talking about everything – it's just layer upon layer, generation upon generation, each one that washes forward, another recedes back into the sea, and there's some more important shells and pebbles on the beach ... I think there's a hunger in all of us, in the human mind to manufacture something greater than itself, to keep itself sane. Sort of, as an animal we need that.

Matthew added:

> That's probably why I do tend to explore the subjects and the religions. I will talk to people, find out what they believe,

because maybe one day I'll find it ... Why don't they invent one I can go to?

RELIGION AND THE PROBLEM OF SUFFERING

Of all traditional beliefs, the assumptions of ethical monotheism are still by far the most prominent, more or less universally present even amongst people who are doubtful of God's existence. They are in no doubt that if there is a God, then God is good, not evil. If there is a God, then God is not capricious, but is utterly trustworthy. Fully in accord with traditional orthodoxy, if there is a God, then God is almighty and all-knowing. These remarkably consistent views can be drawn from almost any of our research conversations.

At the same time, this impressive show of orthodoxy is the source of great ambivalence. God is almost always seen as male and, especially amongst the men we spoke to, punitive. He is the transcendent 'force' who saves and redeems his people, but also condemns them and apparently permits them to suffer. Matthew's scepticism, which I mentioned a moment ago, derives much of its strength from this conundrum. One of the questions in the *Soul of Britain* national survey was put in the following way: 'Some people don't think there is a God. Why do you think this is?' By far the largest single group, 41 per cent of the national sample, agreed with one of the six offered alternative responses that stated: 'There is too much suffering, poverty and injustice in the world for God to exist'. It is often thought that the advent of the scientific attitude is the major stumbling block to religious faith. But the number of people concerned about suffering was almost double the size of the group who felt that loss of belief was because 'science has explained the mysteries of life'.

The figures for the national sample are fully supported by the content of our research conversations. Whilst we have seen how religious experience enabled Mary to cope with great anguish, more than half of those we spoke with brought up the question of suffering as a difficulty for religious belief. When Robert was asked if he believed in God, his reply implied that he was too scared to say 'No', suggesting that the picture in his imagination is of a vengeful God who might punish him for his doubts.

Nicola, who was more disposed to admire traditional belief than most other members of the group, saw the existence of suffering as the main basis of the atheist critique of religion. One of the reasons she gave for hiding her religious opinions from others was because they would very likely trap her by saying: 'Why do you believe in a God that's up there? You never see him, and all these horrible things happen in life'. Simon movingly expressed the implausibility of believing in a good God in his account of his direct meeting with suffering:

> I've had three tragic deaths in my life. There's my mother who committed suicide, then my brother who died of a heart attack in his thirties. As long as I can put logic to things I'm all right. Like my mum, she had an illness so I can live with that. And my brother, who was fit to a certain point but all his life he had [a weak heart] ... I can put logic to that. But like, with my nephew, he died at age two. He strangled to death ... I can't put logic to that.

Jenny talked about how a relative's faith had been rocked by the death of her daughter. She had asked: 'What on earth reason could God have had to take my daughter away at the age of eleven, and make her suffer for two years before she died?' Jenny reflected on 'whether that would be easier if I could just say, "Well, there is no reason. It's just one of those shitty things that happens in life"'.

It is important to add that some believers seemed to see no difficulty over a punishing or vengeful God. For Elizabeth, with her very strong tribal identification with her Scottish Catholic upbringing, God is not only watching over you, he is also 'always behind you'. She had a sense of God permanently spying on her, so to speak, ready to punish her for her sins, which from her very traditional stance she felt was 'fair enough'. Mary, from a not too dissimilar Irish background, had a strong desire for social justice and the alleviation of suffering. Yet her awareness of widespread injustice did not appear to disturb the intensity of her positive feelings about God. She remarked that, 'I don't think I've ever been angry with God for any reason', though in other passages of conversation her memories of childhood images of God were not so benevolent. Sarah was still more ambivalent:

> When I hear of all the sad things that happen and it's sometimes [I] say, 'Oh I don't believe in God', but I still do because I

> still think things happen for a reason. And I feel like a lot of suffering is what we do to ourselves, what we do, what mankind has done to man.

Belinda was distressed because of the funeral, just a week previously, of someone she had known since her schooldays, who had died in a freak accident. Somewhat desperately, she hung on to the belief that there is a purpose behind such apparently random events:

> I believe he has reasons for what he does. I mean you have to hope that he does because [of] some of the terrible things that do happen. And when you lose somebody close to you or somebody dies very young, you have to believe that there has to be some reason why this thing happened. I think otherwise it would be all really too much, wouldn't it, to deal with?

The ambivalence hinted at by Belinda came out fully in what Tracy had to say to her friend Janet:

> She lost a baby and I think I said to her, 'I don't think there is a God, Janet, because how could he let this happen to you. Well, when I do get there, I shall have some questions for him.'

This is not far off from a line in Samuel Beckett's *Endgame*, 'The bastard, He doesn't exist.' Suffering and atrocities make belief hard, and for some people the final nail in the coffin is driven home by the fact that many atrocities are committed on behalf of religious institutions. As Graham said,

> I mean you've only got to look at the world today and a lot of the wars and a lot of the killing is all so-called religion based, isn't it?

The contradictions created for belief in a good God by the existence of suffering can sometimes lead to a denial of orthodox monotheism. Traditional monotheism is still the natural religious assumption for the ordinary person in the street, but as a practical (as opposed to theoretical) belief the idea has been watered down to mean not much more than that God will intervene if my relatives or I get into difficulty. People speak of God helping their families at the same time as having difficulty in believing in God because of all the disasters in the world. Here is Steven (whose wife is a practising

Catholic) talking about homeless people:

> ... this is, this is somewhere where um, God could do more if he could. If there was somebody, um, he'd look after everybody in the world. But, um, if he's that good, then why is there so much suffering? [Does that make it hard for you to believe in God?] Yeah, I think it does. Um, I suppose I believe in him in a, in a small sort of way whereas, um, like he's looking after my children and the children's schools and sort of, um, all the Catholic friends what, you know, um, that we have, that we know. But then again, like I say, there's that much suffering and pain in the world that, maybe, you know, there isn't, there isn't a God.

So God has shrunk down to become a 'household god', looking after an individual's family, but not able to intervene in a wider context. Thus in the case of some people (for example James and Sharon) there is a suggestion that there is more than one God, with the loving household god quite distinct from the remote and frightening creator God, who does not care for his creation.

STEER CLEAR OF THIS AREA

One of the commonest responses to questions about spirituality, or what many people called the 'big questions', was the advice to avoid them altogether. The sense was that the treadmill of multicultural modern life has led to a feeling of being suspended in a fragmented world of 'bits and pieces' from which all coherence has gone. Sarah, who used to think a lot about 'What are we doing here?' had pushed that aside with a decision to 'think positive' and throw herself into ensuring that her children were cared for properly.

A hint of the vague unease surrounding the vacuum in meaning came from Tracy. She had earlier said that her normal response when questions of religion or spirituality came up was to make fun of them. But she also remembered that once in her childhood she had gone into the bathroom and asked her mother what the point of life is if everybody dies in the end. Presumably she did not get a satisfactory reply, for she had long since pushed such questions out of her mind. It was as if now that she was an adult, she had to put away childish things.

To summarise, the best advice many of our informants could give about the big questions was similar to the counsel at the end of Voltaire's *Candide*,[5] 'Don't get morbid.' 'Try not to think about it.' 'Just get on with your life.' This was clearly articulated by Jenny, whose views were almost a reversal of the sceptical Auguste Comte's three-stage view of the growth of understanding – in childhood we are theologians, in adolescence we become metaphysicians, but in maturity we become scientists. For Jenny, metaphysical or religious questions seemed to arise later rather than earlier:

> I think we all go through that 'Why are we here?' Maybe it's just something that comes as you get older, 'Well perhaps I have to accept that there are some questions in life that there are no answers to' ... there doesn't have to be a reason why we exist.

Even Nicola, with her strong religious connections, said,

> I think you just have to make the most of what you've got at the time and not think too deeply about it. Because if you go into it too deeply, ... 'Why am I here?', 'What am I doing here?' it could be a bit negative feeling, if you aren't coming up with the right answers.

In the end, in spite of their assertions, none of those we spoke with succeeded in heeding their own good advice, though some managed to push it out of consciousness for most of the time. In reality no one lives without some theory of why there is something and not nothing. Either implicitly or explicitly, every individual we spoke with had a notion of 'what it's all about'. Very frequently their reflections could properly be called theological, because they concerned the nature and attributes of God. I am even inclined to say that *especially* in the case of those few who asserted that they were atheists, much of their speculation had an existentially committed quality that lifted it above mere sophistry. Spirituality was in no case a dead issue. These questions mattered because they were associated with an awareness that was undeniable, though difficult to express and varying widely in 'range' from person to person.

SOMETHING THERE

For those at the sceptical end of the spectrum of belief, the deficiencies of religious interpretations of reality were felt to be devastating and unanswerable. Yet, set against these rational objections was an undeniable awareness, by far the commonest and most impressive phenomenon we encountered during our research. Underneath the confusing variety of interpretation, ranging from explanations couched in the language of mainstream religious orthodoxy through to bizarre personal speculation, there loomed an all-pervasive sense of 'something there'.

The virtual universality of this experience amongst the members of the group calls for investigation, but its vagueness is problematic. At least since the time of Descartes we have liked our realities to be clear, distinct and sharply focused and this 'something' is manifestly not like that; it is vague to the point that people feel it is impossible to put into words. As a result, whilst the something there was obvious and indisputable, the interpretations discussed earlier in this chapter were almost always presented in a tentative manner. Most people were uneasy about saying *anything* positive about their experience and I have already noted that when they did venture into words their unease was conveyed non-verbally as a sense of bodily tension or embarrassment.

Certainly doubts about the plausibility of religious belief, along with the sheer lack of a religious vocabulary, have both played a part in generating the discomfort felt by our interviewees. This is hardly surprising after more than 400 years of increasingly sophisticated criticism of religious belief within European culture. Especially at the street level of guying and ridicule, no one can remain entirely unaware of the critique. Furthermore as we have seen, most people recognise a connection between religion and spirituality, to the extent of at times considering them synonymous. Hence we might assume that dismissal of one implies dismissal of the other. Nevertheless I am convinced that it is a serious mistake to equate the silence in relation to spirituality with scepticism. The research by Zinnbauer, Lambert and others showing a decline in the idea of a necessary connection between institutional religion and personal spirituality strongly suggests that I am right.

So why are people reticent when it comes to being more specific about 'something there'? People repeatedly tell us that the silence or embarrassment is due to an awareness of the traditional link combined with a suspicion or resentment of religious doctrine seen as a less than adequate interpretation of profundity. The cut-and-dried answers they have been given seem simplistic and deny the mysteriousness of life. This came out particularly devastatingly in James who spoke of that which he encounters as 'deeper than God' and of his communion with this as 'more profound than prayer'. James' disappointment with his theological studies at university related to what he saw as the triviality of the way doctrines were purveyed by his tutors, intellectual playthings divorced from the marrow of life.

Quite apart from the findings of our recent investigation, my experience over thirty years of research is of people repeatedly telling me how the intense meaningfulness of this area of their experience defies expression. Another way of looking at their disquiet at having to name the 'something' is to see it as falling into the *apophatic* tradition in theology. This is the stance that insists that anything we say about transcendent reality is always limited by the language and assumptions of a specific culture. Therefore by definition no humanly constructed language can comprehend the infinite, which can only be referred to obliquely or hinted at by the refusal to make any positive statement.

Inarticulacy in the presence of the sacred or holy has many precedents and is central to the whole project of mystical theology. In the Bible there are repeated references to the nameless and imageless God, represented by the letters YHWH, or, in the New Testament by St Paul's metaphor of 'seeing through a glass darkly'. A strong representation of the same transcendence can be seen in Judaism and in Islam, with their prohibition on any images of God, as also in the practice in orthodox Judaism of writing the divine name as G-d. The apophatic approach is sometimes referred to as the *via negativa* (negative way) and Christian mystics like the Dominican Meister Eckhart have been charged with heresy because of their denial of attributes to God. Another great mystic, St John of the Cross, was accused of being a crypto-Buddhist[6] (that is, an implicit atheist) because of his negative paradoxes as he seeks to approach the all:

> To reach satisfaction in all
> Desire its possession in nothing.
> To come to possess all
> Desire the possession of nothing.
> To arrive at being all
> Desire to be nothing.
> To come to the knowledge of all

> Desire the knowledge of nothing.
>
> ...
>
> When you turn toward something
> you cease to cast yourself on the all.[7]

The well-known Zen Buddhist saying 'If you meet the Buddha on the road, kill him', has a parallel implication, that is, any conscious notion you may find entering your head about the nature of ultimate reality is an illusion that must be destroyed. And in the three-thousand-year-old *Brhadaranyaka Upanishad* we find the earliest mention of the Hindu Vedanta teaching of 'neti, neti' ('not that, not that'), a refusal to apply any description to sacred reality.

It may seem that by associating this esoteric mystical tradition with a group of perfectly ordinary, non-churchgoing citizens of Nottingham I am making an exaggerated claim. The apophatic approach is usually seen as far along the path of development of the devout practitioner of prayer or meditation. Contemplative monks and nuns are expected to reach this stage after many years of ascetical practice, and beginners are warned off mystics like St John of the Cross as 'strong meat' that they will almost certainly misunderstand. Could a group of people who don't even belong to a religious institution possibly have any understanding of the *via negativa*?

I am not claiming that our Nottingham citizens are mystics who have advanced far along the spiritual path. (Although who is to judge? Some of them may have.) But I do think that the intellectual context of contemporary life opens up ways of understanding reality that were very difficult or impossible for our forebears. One of the insights that has become a commonplace in these postmodern times is the notion of the provisionality of labels; the fact that names are arbitrary and therefore have no absolute reality. This is not necessarily self-evident, as is amusingly illustrated by a story told by the Soviet psychologist Lev Vygotsky, of a Russian countryman:

> who said he wasn't surprised that savants with all their instruments could figure out the size of stars and their course – what baffled him was how they found out their names.[8]

In the unconscious meaning-making that goes on as we enter our native culture we are like Vygotsky's Russian peasant, assuming that the words and explanations we are given are unchanging features of the world, almost part of the physical nature of the objects they refer to. That is how it feels in childhood because of the way we learn to speak. Our parents repeatedly associate a particular word with some object, say a chair, until by an astonishing leap of insight we get the idea and learn that that thing over there *is* a chair. Much the same process of associating not only words but meanings goes on when we start school and learn jargon, definitions and technical explanations for subjects ranging from sport to physics.[9] Stability of meaning matters a lot and is especially clung on to by schoolchildren. I well remember feeling resentful when, beginning to study science in senior school, I was told to forget the simplistic scientific explanations I had learned in earlier years. We hang on to those 'sedimented' meanings because it is uncomfortable to be in a state of uncertainty.

The idea that humanly created words and meanings are eternal, divinely given, would have been much more commonly held prior to the Enlightenment because of the dependence on external authorities such as the church or the sovereign for authoritative knowledge. Since the sixteenth century there has been a steady erosion of this view, leading in due course to the relativism of the postmodern movement. Postmodernism has its own paradoxes and difficulties, but it has had at least one important liberating effect. Compared with medieval times there is a much more general recognition that it is human beings who name the stars, just as they calculate how far away they are and speculate on how they came into existence.

Human beings also change their minds and in the process the meanings of words alter or become redundant. It has become clear that whilst there may be truth to be discovered (as most scientists, including myself, believe), at the human level meaning is always provisional. Ordinary people are much more aware of this than their medieval ancestors, so at the level of the fluidity of ascribed meaning they are on a par with a dedicated medieval mystic in recognising the limits of language. That does not of course imply that their practice is even remotely like the state of someone engaged in advanced mystical prayer (though it might be).

SPIRITUALITY AS LOVE

There was one area where our interviewees were not reticent on the subject of spirituality. Quite unprompted by us and more or less universally they insisted on explaining spirituality as another word for disinterested love. There was no difference between people on this issue. Where there were radical differences was in setting a boundary to the 'range' of loving spiritual awareness. At its narrowest, spirituality was limited to a feeling of loving unity with at most one or two other people, whilst at the other end of the scale it comprehended a relationship with all people and all things.

People's first thoughts were concerned with a stable, reliable personal relationship with another human being, usually a partner and/or children. Over and over again, people insisted that their children mattered most to them. Joanne, when she was asked for a significant moment in defining her life said, after a long pause,

> When I became a mother was a very significant moment … your whole life changes absolutely completely in a few seconds … For the first time in my life I felt I'd got something to love that really loved me back … I really felt totally in love with this baby and this baby needed me. I never felt that my parents particularly loved me …

In quite a number of cases the aspiration towards loving relationship stopped short at this first level, the boundary of the immediate family. It was as if there was an invisible protective wall surrounding the home. Tracy was still in her twenties but was deeply pessimistic about the human condition. She felt that social cohesion had broken down so much that the only stability to be found was inside the house:

> I think the only religion we've got is ourselves, isn't it? You know, believe in yourself and you get on with life; it's such a madhouse. You just have to get on and look out for each other.

Colin made his living as a manual labourer and saw his life as precarious, 'surviving from day to day, doing the best you can'. His strategy was to stay close to the family unit: 'We are a tight knit family. We see our relations every so often. Apart from that we just keep ourselves to ourselves.' As for the outside world:

> You see the news and what's happening in the world – you kind of think well it's not happening here … you see all these kidnappings of children and so forth but if it's not happening near here you don't tend to worry about it … we never even tend to

> talk anything about ... what's happening in the world.

Lucy's sense of spirituality as relationship reached somewhat further. She saw a connection between the bonds of affection that were important to her and what some people get out of religion. The implication was that the latter were caught in an illusory relationship with a non-existent God.

> My life revolves around those relationships, and when am I going to see them next, and what are we going [to do] together, and it's just that feeling that you get when you're with someone that you know and love. And to me it's a real buzz. I can imagine that some people get that through religion, but I get that through real people.

Joan was not so scathing about the religious institution. She felt it was the loving links within the family and, beyond that, the religious community to which she half wanted to return, that gave people security. Debbie also saw it as important to spread her love. She looked after people who were dying and liked her job very much. She was aware that there was a spiritual dimension to her vocation that would make it difficult 'if I wasn't sort of religious in any way' ... The good thing about her job is:

> ... giving people the comfort really. You almost feel like you're an angel going in. Just holding somebody's hand or being there for them goes a long way. And they seem to see you as some sort of healer sometimes. I feel I'm getting a lot of pleasure out of it, a lot of comfort as well.

Mary, who apart from non-attendance at church was conventionally religious, had a heroic view of the nature of the loving relationship. Her care for others took it well beyond the immediate family, or even the local community, into worldwide concern within the embrace of a loving God, as she saw it. Indeed one of her major criticisms of the religious institution from which she had withdrawn was its narrow understanding of the vocation of universal love:

> I'm talking about humanity on the whole, really. I mean it's all very well raising money for the poor and feeding them when there's a disaster or something like that. But this isn't just for the church, this is for politics, this is for all of us. These people

> have been through a famine or war or floods ... and we all dip our hands in our pockets and feel holy and we think that's that. But these people remain in this situation.

The association of spiritual insight with loving service is a reminder of the traditional view that there is a close connection between ethical behaviour and ... what? Religion? Spirituality? Once again we are faced with the question of whether there is some identifiable quality that links all spiritual awareness and how this ties in with religion. During the 1990s, Rebecca Nye and I tried to find out, and what we discovered is discussed in the next chapter.

Chapter 6

PRIMORDIAL SPIRITUALITY

And even if this term [God] were ever to be forgotten, even then in the decisive moments of our lives we should still be constantly encompassed by this nameless mystery of our existence … Even supposing that those realities which we call religions … were totally to disappear … The transcendentality inherent in human life is such that we would still reach out towards that mystery which lies outside our control.

Karl Rahner[1]

For babies, every day is first love in Paris.

Alison Gopnik[2]

THE SOURCES OF WHO WE ARE

Who can possibly doubt that culture plays a major part in making us into the people that we are? My earliest memory of being literate is of standing by my mother's knee, reading to her out of the Bible. From a very young age I was encouraged to learn off long passages from Holy Scripture by heart, with the result that I can still give an accurate rendering of many of them to this day. Furthermore, their association with the unreflectiveness of early childhood means that they continue to carry an extraordinary amount of emotional weight.[3] The language and imagery of the Bible were so pervasive that they coloured my interpretation of the landscape. The Northern Scottish village in which I grew up lies at the base of a heather-covered extinct volcano, down the slopes of which, in my infant mind, Moses had stumbled bringing with him the Tablets of the Law. Jesus was baptised in a trout pool in the burn at the bottom of our road and, one warm summer evening, two disciples walking out of our village felt their hearts burn within them when they encountered the risen Christ on the road to Emmaus.

There must have been many people in the village who didn't go to church, but as a child I either ignored them or was unaware of them. In practice I lived in a monoculture, only marginally intruded upon by alternative ways of life. This is true of any society in which most or all of its members share a set of beliefs and values (including of course secularist beliefs). Given the obvious importance of socialisation, it is not hard to believe that the sources of the self[4] are to be found entirely in the details of a person's life history. Since the self, at least in the case of religious believers, includes a supposedly spiritual dimension, that too must be 'socially constructed'. In what sense then, could spirituality be said to have more than the most fragile basis in reality? The difficulty is that if we accept the social construction argument, there is a severe problem associated with the fact that the process of socialisation begins in childhood. After all, kids will believe anything.

We surely need to be wary of the credulity of children and of the charge that certain kinds of child-rearing amount to an indoctrination that lasts into adult life and blunts common sense. Richard Dawkins as both biologist and critic of religion puts this forcefully:

> Natural selection builds child brains with a tendency to believe whatever their parents and tribal elders tell them. And this very quality automatically makes them vulnerable to infection by mind viruses. For excellent survival reasons, child brains need to trust parents and trust elders whom their parents tell them to trust. An automatic consequence is that the 'truster' has no way of distinguishing good advice from bad. The child cannot tell that 'if you swim in the river you'll be eaten by crocodiles' is good advice but 'if you don't sacrifice a goat at the time of the full moon, the crops will fail' is bad advice. They both sound the same. Both are advice from a trusted source, and both are delivered with a solemn earnestness that commands respect and demands obedience.[5]

Dawkins' description of religious ideas as 'viruses of the mind', or an infection,[6] implies that unlike scientific ideas, they have no other empirical basis than social transmission. This is a view that carries over rather easily into the notion that the spiritual life is also entirely a matter of induction into a culture, because of its historical link with religion. It is an opinion that is very widely held nowadays, sometimes (minus the pejorative reference to 'virus infections') by theologians.[7] Granted that there is a great deal of truth in this way of accounting for the content of spiritual experience, it is nevertheless implausible to attribute it to *nothing but* social construction and it is slightly surprising that a fellow biologist allows such dominance to metaphorical viruses in the mind. The sociological argument omits the rather obvious physical fact that experience is not an isolated 'something' that is inscribed onto people's awareness as if they were entirely passive recipients. We are embodied beings with a genetically inherited physique, temperament and subtle array of inbuilt competences, including awareness or perception.

This ought to be uncontroversial, but the sociological account of who we are is so dominant that the case for interaction needs to be argued more thoroughly. Back in the 1980s Charles Lumsden, a medical scientist, and Edward O. Wilson the founding father of sociobiology, devised a story to help us to see why this must be so.[8] They asked their readers to imagine a distant planet inhabited by an alien species named *Xenidris anceps* ('Resting place of indecision'), whose members have minds that are blank slates. Their genes control their physical structure, including their brains, but their mental life and behaviour are entirely free, going in every conceivable direction according to the accidents of their life history. Many social scientists and advocates of an extreme version of postmodern relativism implicitly take for granted a 'sociologism' that amounts to saying that human beings are like the xenidrins. Lumsden and Wilson offer the example of the anthropologist Leslie White, who explicitly claimed that culture submerges the effects of biology 'to the point of insignificance.'[9]

In case you think that I am about to go to the other extreme and suggest that spirituality is entirely a matter of biology, let me complete Lumsden and Wilson's story. On another distant planet live aliens who are genetic robots with the scientific name *Eidylus strictus* or 'rigid experts':

> The eidylons are a brilliant and formidable species, but their entire thought and behavior are programmed into their brains, right down to the exact words they use to piece together sen-

> tences. Language, art, and every other aspect of the culture are affected by the circumstances of the moment, but all are genetically predetermined in form ... and here is the paradox that illumines the problem of mind and intelligence; the eidylons also teach their young all the details of eidylon culture. How can a civilisation be simultaneously genetically fixed and transmitted by culture? The answer is that although the eidylons teach and learn all that they know, they can transmit only one thing in each category – one language, one creation myth, one set of hymns and so forth. Like the white-crowned sparrows of California (Earth) which must hear the song of their own species in order to learn it but are impervious to all else, they are unable to learn anything but the eidylonic repertory.[10]

Unlike Wilson, who has been unfairly accused of 'geneticism', some biologists used to argue that human culture is largely genetically determined in this way. Most notoriously, the Oxford botanist C. D. Darlington suggested that amongst other things genetics determines the incest taboo, belief systems, castes, skills, trades, moral codes, militarism, pastoralism, social classes, being a prophet, horsemanship, ability to govern, Christian tolerance, Jewish religious devotion, homosexuality and the master-slave relationship.[11] If Darlington were proved to be right, then social factors rather than biological factors would be insignificant in the construction of the self.

In summary, two simplistic ways of explaining the passing-on of experience from one generation to the next are via cultural transmission (the xenidrin way), or by genetic transmission (the eidylon way). It should be clear that both these extremes are non-starters. Whatever the self may be, it comes into existence through a continuous interaction between genetics and environment. Although in every human group there is a vast array of cultural possibilities, genetics in the end sets limits. And here we are in fact returning to an updated version of Alister Hardy's ideas on 'behavioural selection'. Lumsden and Wilson put it this way:

> Even if a species could be created to resemble the xenidrins, with pure cultural transmission, evolution will eventually carry it from the blank slate and into a culture based on gene-culture transmission ... It is inevitable that some of the choices available to the society (say in diet, sexual behavior, or ways of

counting) will confer greater survival and reproductive ability on the members who make one choice and spurn another.[12]

Hardy would have wanted to add the choice of attending to spiritual awareness, because it confers survival value. Since the publication of *Promethean Fire* a whole new science of *memetics* has developed, referring to cultural transmission. It takes its name from Dawkins' term *meme*, coined in 1976 to mean a new kind of replicator, 'a unit of cultural transmission, or a unit of *imitation*'.[13] Memes might include any unit of human behaviour that can be imitated and Dawkins gives many vivid examples including 'tunes, ideas, catch-phrases, clothes, fashions, ways of making pots or building arches' as well as scientific and religious ideas (these last being the memes he likes to call viruses). The idea of the meme is still highly controversial,[14] but its prominence does at least imply an increasing acceptance that the self originates from both biology and culture. The problem is that memes are commonly discussed as if they were alien invaders, completely isolated from physiological constraint, and as we shall see in Chapter 11, this is very far from the case.[15]

PRIMORDIAL SPIRITUALITY?

Turning now to physiology, can we discover the nature of the inbuilt awareness that is involved in spirituality? Whatever it is it certainly has emotional power. For some years I was on the staff of the Centre for the Study of Human Relations in Nottingham University. I sometimes used a very simple exercise with students to help them to explore the question of identity. I invited them to form pairs and take it in turn to ask their partner repeatedly and quietly, 'Who are you?' The early responses to the question are usually clichés to do with social attribution and positioning. 'I'm David Hay', 'I'm a Scotsman', 'I'm a married man with three sons', 'I'm a zoologist' etc. After a while the attributions dry up and there is a silence, a period of boredom, and (if the person is prepared to persist) a move into deeper and more puzzling territory. After all, who is this nameless naked body that by an accident of history is given a name, a nationality, a social position and a covering of clothing? It is typical of the 'Who are you?' exercise that it is moving to the participants, quite often leading them to weep at what they experience as the profound strangeness of the human condition.

They weep because they are entering existential depths shared by every human being. They are depths that were thoroughly explored in our Western culture 1,600 years ago by St Augustine in his *Confessions.*[16] The *Confessions* are written entirely in the form of a prayer, an extended contemplation on his life by Augustine who, as a Christian bishop, chooses to do so in the presence of God. As the 'Who are you?' exercise progresses, many people – even those who assert that they are without religious belief – find that their mood similarly becomes more and more contemplative. Perhaps when we find ourselves in this territory it is impossible *not* to be contemplative. There is the sense here that we are, whether we like it or not, spiritual animals. Thus, the Jesuit theologian Karl Rahner, in one of his reflections reproduced as an epigraph to this chapter, invites us to imagine a world that has become so deeply secularised that the word 'God' has disappeared from the dictionary. Even then, we would still find ourselves embraced by the fundamental mystery, 'Why is there something and not nothing?'

To gaze directly at this question is not to wallow happily in ignorance, but to explore the mystery of Being using the method most appropriate to it – observation. During the 1990s, with the help of my colleague Rebecca Nye, I decided to take this exploration further by setting up a three-year investigation of the spirituality of young children. Accepting for the time being the metaphor of the meme, I reasoned that the array of memes constructing the ideology of secularism tends to lump both spirituality and religion together as wrongheaded. Hence any naturally occurring spiritual awareness in Western populations would be most easily found in children. At that age they are less likely to have closed their minds to the premise that the spiritual dimension of life is real and meaningful.

In an oblique fashion I was supported in my assumption about childhood by two items of information, one small, the other very large. Several years previously, when I first began investigating these matters, I was alerted to the issue by a conversation with a young zoology graduate who explained to me that when she was younger she had often been aware of a spiritual dimension to her experience. On reaching the senior school and beginning a serious study of science she realised that 'that sort of thing can't happen' and indeed it stopped happening.

The second, much larger piece of *a priori* evidence came from research findings published in 1991 by the Finnish psychologist of religion, Kalevi Tamminen at the University of Helsinki. Tamminen's book *Religious Development in Childhood and Youth*[17] includes an investigation of the spiritual experience of a sample of over 1,300 Finnish children and young people drawn from seven widely scattered communities in the southern part of the country. They ranged from seven to 20 years of age. Tamminen used two basic questions. The first was 'Have you at times felt that God is particularly close to you?', followed by an invitation to talk about the experience and the circumstances in which it took place. The second question asked 'Have you at times felt that God is guiding, directing your life?', again with an invitation to describe the experience. In addition, Tamminen showed his subjects photographs of children in religious contexts as a means of eliciting 'projective' responses. Thus one photograph showed children attending a church service, with a caption beside it that read, 'Henry and Lisbet went to church. They are sitting in church now. How are they feeling in church ... What are they thinking?' The assumption was that the answers to the questions would express something of the feelings of the respondents, were they to be in a similar situation. Tamminen found very high levels of report of the nearness of God (66–76 per cent) in the case of children between the ages of seven and 13 years, followed by a very marked drop in the older age range. In a longitudinal study a group of children were interviewed at the age of nine, then re-interviewed at 11 and 16. Amongst the 16-year-olds, frequent experiences of God's nearness had dropped to less than a third of what they had been when the same pupils were aged nine.

It occurred to me that the point at which report of the presence of God took a sharp turn down coincided with the age at which most children are inducted into the sceptical tradition of the Enlightenment, particularly via their scientific education. As the young zoologist I mentioned a moment ago had told me, according to that tradition such experience can't happen. The established secularist view of this shift is that it is the result of a growing rationality as the child matures intellectually.

I mentioned Auguste Comte briefly in relation to Jenny. Comte first published his *Law of Three Stages* at the beginning of the nineteenth century:

> From the study of the development of human intelligence, in all directions, and through all times, the discovery arises of a great fundamental law, to which it is necessarily subject, and which has a solid foundation of proof, both in the facts of our organisation and in our historical experience. The law is this: that each of our conceptions – each branch of our knowledge – passes successively through three different theoretical conditions: the Theological or fictitious; the Metaphysical, or abstract; and the Scientific or positive.[18]

Comte saw this sequence in the unfolding of European history and also recapitulated in the life of each individual; theologian as infant, metaphysician as a teenager and scientist in maturity. In effect, we tell fairy tales to gullible children, but once they reach the age of reason they abandon them. But could it be that Comte was himself trapped by the intellectual assumptions of the eighteenth-century French *philosophes* who were his mentors? What if it is the ideology of secularism that overlays and suppresses an original biologically inbuilt spiritual awareness present in all children? What if the children Tamminen spoke with were telling the truth about their experience and not merely parroting religious clichés taught them by their parents and teachers? How could we try to find out more about its nature?

On the basis of these ponderings, I decided to investigate how British children at six and ten years of age would respond in conversations about spirituality. In some ways the study would be like Tamminen's Finnish programme, but with certain modifications to broaden its scope. Like almost all previous researchers in this field, Tamminen assumed that spirituality would express itself through Christian language and imagery. But if spiritual awareness is natural to us, then it is likely to manifest itself in other ways, especially in the secularised West. Rebecca Nye and I had to devise ways of getting in touch with this dimension of experience without using formally Christian language, or for that matter any religious language, because it would most likely restrict what might be perfectly valid responses to our enquiries.

But where could we begin? What we were doing was so new to us that it felt like starting without a starting point. In the words of an old joke, we would be wise not to start from here. At first we thought we might get started on the basis of a clear definition of the word 'spirituality' obtained via discussions and seminars with academic colleagues who were knowledgeable about the subject. Unfortunately the variety of opinions that emerged simply increased our confusion, but something interesting did emerge from this apparent failure. Everybody agreed that whilst pinning spirituality down to an agreed definition seemed impossible, they were able to recognise it when they came face to face with it. Taking that as our cue, we spent a considerable amount of time identifying some areas of ordinary life where, if there is such a thing as spiritual experience, it might be expected to turn up. We settled on these areas not only because they are associated with the spiritual life, but also because they are prominent aspects of childhood. I'm sure there are many such areas, but our initial tally was three, all connected with awareness: Awareness of the Here-and-Now, Awareness of Mystery and Awareness of Value.[19]

AWARENESS OF THE HERE-AND-NOW

I first picked up this idea from noticing the emphasis given in several forms of Buddhist meditation to keeping one's awareness in the present moment. Often this is done by observing with great attentiveness the rise and fall of the abdomen during the breathing, as it happens, or by noting the fine detail involved in taking each step as one walks back and forth very slowly.[20] In Theravadin *vipassana* meditation, literally 'awareness' meditation, the practice leads towards an immediate attentiveness to all that arises in consciousness, whether feelings, thoughts, painful or delightful memories, even the sense that the exercise itself is a useless waste of time (whilst briefly noting the feeling of uselessness as the current content of the awareness, then quietly returning to the breathing or walking).[21] Strict Theravada doctrine refuses to make any statement for or against the existence of God, and for that reason is often labelled as atheist. Yet when we look closely at the devotional practices of the theistic religions, they lay exactly the same emphasis on what the eighteenth-century French Jesuit Jean-Pierre de Caussade called 'the sacrament of the present moment'. That moment, says Caussade, is the point where one is most clearly and directly in the presence of God.[22] Thus novices are taught that prayer is the raising of the heart and mind to God, here and now. If it ceases to be that, for example by thinking *about* God rather than simply remaining in the divine presence, strictly speaking it is no longer prayer; it has turned into theology.

To maintain awareness of the here-and-now is to return to a kind of awareness found universally in infants. The Edinburgh psychologist Margaret Donaldson[23] calls this state the 'point mode' and notes that babies remain quite strongly in this mode until about the age of eighteen months. Far from seeing this as a deficit, students of child development are coming to suspect that 'babies and young children are actually more conscious, more vividly aware of their external world and internal life than adults are'.[24] This might suggest that meditation and contemplative prayer are attempts on the part of adults to retrieve the vivid here-and-now awareness of the infant. Beginning at about eight months infants gradually and progressively enter the 'line mode' in which more and more time is spent dwelling on the past or the future, regretting past errors, enjoying delightful memories, planning, hoping or fearing what the future will bring. In adult life the line mode is so dominant that the here-and-now may play only a minimal role. Since young children are naturally very close to the mode associated with the practical exercises of the spiritual life, we thought this was an area of their experience worth exploring.

AWARENESS OF MYSTERY

As I hinted a moment ago, childhood is often characterised as a time of deficit. Children are seen as inefficient adults who are physically weak and must develop strength; ignorant and must be given understanding so that they can enter society and earn a living. The process of formal education through which every Westerner passes is designed to make up for these weaknesses, especially to replace ignorance with reliable knowledge. And it is true that by the time a person has reached the end of their secondary schooling, many of the puzzles that seemed inexplicable in infancy have been given intellectually satisfactory explanations. Quite commonly in late adolescence young people assume that in principle, if not yet in fact, everything has a scientific explanation. When, like Immanuel Kant, they are amazed by the starry heavens and the human mind, it is with the qualification that cosmologists and neurophysiologists are well on the way to achieving a technical understanding of those remarkable phenomena.

I have already begged to doubt whether all of early life is necessarily to be construed as deficiency. Unless they are ill, children display an alertness and curiosity that eventuates in endless questioning of adults on every conceivable theme, ranging through technical puzzles such as why the sky is blue, or why water comes out of a tap. They also ask other kinds of questions of a type that Comte assured us we would outgrow, such as 'Where was I before I was born?' Schooling certainly provides us with technical answers, but technical knowledge does not necessarily assuage a deeper longing to probe the existential questions that are characteristic of our early childhood. In spite of Comte and his modern successors, a substantial number of adults find that they are unable to leave such ponderings behind and not all of these are mentally retarded. Martin Heidegger, arguably one of the most innovative and influential thinkers of the twentieth century, accused the whole of the Western philosophical tradition of forgetfulness of Being with a capital 'B'.[25] Many adults, including philosophers, suffer from amnesia over every small child's question 'Why is there something and not nothing?' Intuitively Rebecca and I felt that mysteries of this kind fall into the category of spirituality. This is evident in the fact that they are the kinds of questions that all religions attempt to grapple with through a multitude of metaphors, myths and legends. Hence children's endless curiosity suggests that they are often more likely than adults to be in touch with the spiritual dimension.

AWARENESS OF VALUE

What in the end matters most of all to us? It is characteristic of religion that it claims to be concerned with what is more important than anything else. People sometimes die rather than give up their religious beliefs, as attested by the long list of martyrs commemorated in the Christian Calendar. The Protestant martyr Thomas Cranmer, who was burned at the stake in Broad Street in Oxford in 1556, had at first written a letter of submission to the Catholic Queen Mary. When the fire was lit around his feet, as a sign of his repudiation he leaned forward and held the hand that had signed the letter in the flames until it was charred to a stump. Bodhidharma, twenty-eighth patriarch of Indian Buddhism (470–543 CE) is credited with bringing Zen Buddhism to China, where his most eminent disciple and successor was Hui Ko. To this day, Zen novices are instructed in the need for commitment by the story of Hui Ko's search for spiritual enlightenment. He had repeatedly pleaded with Bodhidharma to take him on as a student, but was not admitted until finally in desperation he cut off his left hand as a sign of his sincerity.

An old Zen story reminds us not only of sincerity but the earthiness and immediacy of spirituality. A master asks a young monk who is well advanced in his study of the Buddhist sutras, what is the most important thing of all. 'To follow the Buddha' he says. In response the master plunges the young man's head into a trough of water. He comes up gasping. Again he is asked what is the most important thing of all. 'To understand the Four Noble Truths'. Once more his head is held under the water. What is the most important thing? 'Enlightenment', he shouts. Again his head goes under the water and he comes up choking, fearing he will drown. What is the most important thing? 'To be able to breathe' he screams, and in that moment becomes enlightened. All spiritual questions have this quality of intense immediacy and again we notice that at least from some spiritual perspectives, there are dimensions of immediate experience that are equally or more important than theoretical beliefs. Compared with adults, children express their sense of value in the vividness of their everyday experience of delight or desolation, laughing and crying easily in ways that disappear as we reach the years of discretion. This immediacy, directness and intensity of feeling, so characteristic of spirituality, suggested to us that we ought to include awareness of value as one of the categories.

Having identified what we felt were three important categories of spiritual experience, we then had to devise a way of encouraging children to talk about these dimensions of their lives. Furthermore we wanted to do it without making any prior assumptions about the religious beliefs of the children. The main way we did this was to select a set of pictures[26] of children in situations where spiritual awareness might be expected to be triggered, whilst at the same time avoiding explicitly religious contexts of the type employed by Tamminen in his Finnish study. The pictures included:

- A little girl sitting by the fire in the quiet of the evening, staring intently into the embers.
- A boy in his bedroom. Instead of going to sleep he is sitting up, gazing out of the window at a star-filled sky.

- A girl tearfully discovering that her pet gerbil is dead in its cage.
- A boy standing on a wet pavement looking upwards with his arms spread apart. He has dropped his packed lunch on the pavement and seems to be saying something like 'Why me?'
- A boy standing by himself with bowed head in a schoolyard. The other children are running around the yard, ignoring him.

Rebecca spent much of her time visiting schools with her tape recorder and talking individually with the children, inviting them to have their say on the spiritual life. She was careful to avoid using religious language unless the children chose to introduce it and this is where the photographs were very useful. For example she might show the child she was talking with the picture of the boy gazing at the stars and invite a response with questions like, 'I wonder what he is thinking as he looks up at the sky?' This turned out to be a richly rewarding way to encourage the children to talk about themselves, for more often than not they were imagining themselves into the situation and describing what their experience would be in similar circumstances.

CONTRASTING SPIRITUALITIES

Rebecca's visits to the schools were not all plain sailing. Occasional uncomfortable incidents were a reminder that the research we were doing was controversial and not approved of by everyone. One day she happened to be sitting in the staff room waiting to begin a series of conversations, when a teacher asked her why she was visiting the school. When she explained the nature of her research he became angry and tried to have her thrown out. In spite of difficulties, by the time she completed the fieldwork in primary schools in Nottingham and Birmingham, she had gathered a unique collection of one to one conversations with children about spirituality. When transcribed, they amounted to over a thousand pages of text.

On reading through the whole set of transcriptions what stood out as a vivid conclusion was the undeniable fact that none of the children Rebecca spoke with was without a personal spirituality. Of course what they said was extremely varied, relating to their personalities and home background. Rebecca referred to their individuality by speaking of the unique 'signature' of each child. This ranged between children whose conversation was almost entirely conventional in tone and unselfconsciously religious in content, to others for whom religion was already remote. Yet the latter group quite obviously were not out of touch with the spiritual dimension of their lives. To illustrate briefly the range of response I have picked out two of the children with whom Rebecca spoke to give a flavour of what they said.[27]

At one end of the continuum she talked with Ruth, a six year old, characterised by her as a happy, stable child from a religiously practising home. She went to Sunday School regularly but was very frank about it as uninspiring, telling Rebecca, 'somehow I never want to go, because it is so boring'. But very early on in her conversation it became clear that she had a quite exceptional sense of wonder, particularly in relation to the natural world. Her delight with life bubbled out in one of her remarks to Rebecca while she was drawing, 'I like nature … '. [Why?]. ' … just because I like it. I don't know. And it's so beautiful to be in the world.' Love of nature quite naturally led her to thoughts of God, 'When I see um … the trees burst into life. In spring I like that. But when I see the lambs in Wales, oh … it makes me … oh … leap and jump too!' Ruth's religious imagination struck both Rebecca and me as quite remarkable. Here she is, describing her image of heaven (remember, she is only six):

> A mist of perfume, with gold walls, and a rainbow stretched over God's throne … but a transparent mist, like a … I can't explain it. Like a smell. A real cloud of smell, a lovely smell … like the smell that you get when you wake up on a dull winter morning, and then when you go to sleep, and you wake up, the birds are chirping, and the last drops of snow are melting away, and the treetops, shimmering in the breeze, and it's a spring morning … [then she added] I suppose it's not a season at all, not really, because just a day in delight, every day.

A sceptic might claim that Ruth is being misled on religious matters, but could hardly dispute the reality of her spiritual life. It is overt and expressed in conventional metaphor. But what about the spirituality of ten-year-old Tim, whom we picked out as an example of someone lying at the opposite end of the religious spectrum? The pattern of socialisation within Tim's family led him to refuse any kind of religious label, hence it would have been entirely false to attempt to explore this question with the kind of traditional religious terminology used so unselfconsciously by Ruth. In spite of that, a great deal of what Tim expressed spontaneously in response to the set of photographs was broadly on themes having religious associations: animal reincarnation, the possibility of an afterlife, and the sources of morality and free will, and whether there are many gods or only one. Thus he pondered:

> I sometimes think about if there is one God and there is … everybody, well … most people believe in one God and um … there's um … different people believe in different gods. Which God's real? Um … I just can't figure that out. And I sometimes think about after the universe, what's … uh … what's the universe um … going on for ever. I just don't know. [What does that feel like?] Well when I'm thinking about the universe that gets me quite annoyed sometimes because I can never think about um … get the right answer or get even near it and um … then well … things … you just wonder.

Given the secularised nature of his family environment it is hardly surprising that the distinctive characteristic of Tim's reflections on spirituality was one of inner conflict. In contrast to the immediacy and poetic tranquillity of Ruth's experience, his personal signature, in Rebecca's phrase, was a set of 'conflicting hypotheses representing a special kind of mental work'. On the other hand he resembled Ruth and other children in that his personal style was spread across many other contexts. He also indirectly offered a critique of Ruth's unquestioning religiosity when commenting on the possibility of children having religious experiences such as a sense of the presence of God:

> I think they just look at it and think WOW and uh … forget about it really … Or just um … think about it, but don't think how they were made [i.e. the socially constructed basis for giving a religious attribution to the experience].

Tim presented himself as possibly the most secularised of the children with whom Rebecca spoke during the course of her fieldwork, but it would be a mistake to discount Tim's spirituality on the grounds that his language is strongly sceptical. He was perfectly conscious of the realm being discussed and rather like the young biology graduate I mentioned earlier, could give instances from his own experience, especially when he was younger. His ability to stand back reflectively is related to the fact that he is four years older than Ruth. His dismissal of (at least some) spiritual experience as illusory could easily have belonged to the rhetoric of discernment used by a competent spiritual director within orthodox Christianity. A close examination of the total transcript of his comments points to a powerful spiritual awareness battling with a lengthy cultural tradition that dismisses his searching as a wild goose chase.

DISCERNING SPIRITUALITY THROUGH DIFFERENT KINDS OF DATA

Reading through the transcripts of the research conversations it became clear to Rebecca that the passages indicative of children's spirituality were of two kinds, those using explicitly religious language and others where the spiritual content had a more implicit, hidden quality. Much of Ruth's conversation was in the first, religious mode, as was that of six-year-old John, here speaking very directly of his experience of God:

> Sometimes I feel that ... um ... I am in um ... a place with God in heaven and I'm talking to him ... And um ... there's room for us all in God. He ... God's ... Well ... he is ... in all of is ... He's everything that's around us. He's that microphone ... He's that book. He's even ... He's sticks ... He's paint ... He's everything ... Around us ... Inside our heart ... Heaven.

John abruptly broke off at this point to ask Rebecca if she had seen the film *Indiana Jones and the Temple of Doom*, possibly because he had become embarrassed by his self-revelation. Certainly many children made it clear that such explicitness invited derision, even suggesting that they themselves would probably join in the ridicule if one of their classmates was injudicious enough to speak of their experience in public.

It is therefore not surprising that much spiritual discourse is implicit; that is, expressed in languages other than religious.

Many passages in the children's conversations contained no traditional religious language at all, yet we could not fail to be impressed by their close parallels to the easily recognisable descriptions given above. Sometimes the connection between religious and non-religious language was overt. One helpful example came from six-year-old Freddie, who had much to say about the importance of friendship and reconciliation in his life. The way he expressed himself suggested a depth of importance that strayed beyond the personal into a universal concern – he seemed to be expressing his world-view and, by extension, his 'spirituality'. In Freddie's case he made several explicit links between this and a more conventionally religious expression of his spiritual life. Thus in one conversation about the importance of friendship and his difficulties with maintaining good relationships, he told Rebecca,

> God's the kindest person I know, I think ... because he never shouts or tells you off, because he never even speaks to you apart from perhaps when you're dead.

Another instance of this non-religious expression of spirituality came from Rebecca's conversation with Jackie, who was also six. She was quite overt in her indifference to the formal religion she encountered at school, for example not participating in prayers during morning assembly. Even so she spoke of a 'nice feeling' simply through being present, 'like it seems like ... um ... no unkindness'.

IDENTIFYING THE PRIMORDIAL CORE

I have summarised very briefly a lengthy research procedure that, as it unfolded, engendered a powerful and cumulative sense of being presented with a common universe of human experience. Quite often it was expressed in conventional religious rhetoric, but many terminologies were used, including the languages of science, of fairy tales, of science fiction, of technology, philosophy and still other modes.[28] There were no children from whom this dimension was entirely absent; though in the case of one or two of the ten year olds it was pretty thin. The children themselves had suggested why this might be so. An unsympathetic social milieu required its suppression; and as we saw in earlier chapters, by the time adult life is reached this is overwhelmingly the case. Nevertheless, whether presented in religious or secular language, the common ground was unmistakable and of course on the basis of Hardy's biological hypothesis, is what might be expected.

The close emotional and intellectual parallels apparent in all the material, plus my conviction that spiritual awareness is a general human competence led me to wonder about its primordial nature. Would it be possible to do a textual analysis of the collected children's talk to make clearer what this precursor might be? The job was done with the help of NUD*IST, a computer programme originally developed by Lyn and Tom Richards at La Trobe University in Melbourne.[29] NUD*IST speeds up the process of sorting and interpreting units of meaning in a text, but as we had over a thousand pages of text to go through there was still plenty of work. For months on end the walls of my office were covered with sheets of paper filled with the units of meaning that emerged from Rebecca's laborious line-by-line analysis. Eventually, by repeated inspection of the material, she identified a core category that satisfactorily drew together the relevant data. The term that seemed to fit best was *relational consciousness*, described by Rebecca as consisting of:

- An unusual level of *consciousness* or perceptiveness, relative to other passages of conversation spoken by that child.
- Conversation expressed in a context of how the child *related* to the material world, themselves, other people, and God.[30]

I must admit that I was disconcerted by relational consciousness to the point of being rather disoriented, which I suppose is one of the salutary results of empirical research. Like every investigator, I began this programme with my own set of prejudices about the subject of our enquiry. My image of the true adept of the spiritual life was someone sitting in the lotus position in hermit-like isolation, probably halfway up a mountain in the Himalayas. I thought I had good grounds for my opinion, basing it on scriptural evidence from the great world religions. Thus the *Bhagavadgita* advises, 'Day after day, let the yogi practise harmony of the soul, in a secret place, in deep solitude'.[31] In his Sermon on the Mount, Jesus taught his followers, 'When you pray, go to your private room and, when you have shut your door, pray to your Father who is in that secret place.'[32]

Once I thought about it a little more, the relational aspect of spirituality made eminent sense. I now see the solitude that is so characteristic of much (but not by any means all) prayer and meditation as a setting that is particularly conducive to maintaining the here-and-now immediacy that constitutes spiritual awareness. Jesus made it quite explicit that the reason one should retire to one's private room for prayer is to get away from the kind of temptations to hypocrisy that happen when, for example, members of a religiously devout community set themselves up to pray in public, or give to charity with a trumpet blast; and all the other accoutrements masquerading as prayer that are mentioned in the Gospel. As Jesus said, these people have already had their reward. Privacy here helps people to remain honestly in the immediacy of their here-and-now relationship with reality. Relational consciousness is a pretty good name for the natural awareness that Alister Hardy had in mind. It is because we have this awareness that we can be spiritual in the first place. It allows the possibility of relationship to God, or, if we are non-theists, the sense of a seamless 'holistic'[33] relation to the Other, whether conceived of in secular or religious terms.[34]

The reference to holistic relationship is a reminder of the 'point mode' of infancy and of our universal beginnings as human beings. We body forth from nothingness into an intense relationship, enclosed in the warmth of another person for three quarters of a year. No greater intimacy is possible. For our mother it is a profoundly physical experience, changing her biochemistry, her shape, the way she walks, as well as the thoughts and feelings that fill her whilst we are developing inside her uterus. At a certain stage we make our presence felt by kicks against the walls of her abdomen that are perceptible to an outsider. In all probability she talks to us, expressing her feelings about the relationship. Then, approximately nine months after the process began, we have the experience of moving out of the enclosed universe of her body and into yet another world. It is the normal (as opposed to abnormal, nightmarish) experience, that the intimacy is not broken at birth. When we emerged from our mother's body it is highly likely that we were looked at intently, cradled, embraced, stroked, washed and fed.

It is not always noticed that newborn infants are anything but passive recipients of all this loving attention. Back in 1973 when Olga Maratos submitted her PhD thesis on imitation in the first six months of life to the University of Geneva,[35] she was struggling against the currently dominant scientific orthodoxy. She was the first person to present scientific data supporting what mothers have always known in an informal way; that you can have a conversation with an infant. The great developmental psychologists of the time, people like Jean Piaget and B. F. Skinner, had insisted that intelligent exchanges between infants and adults were impossible at such a young age. Subsequent research on infancy and early childhood has shown they were seriously mistaken.[36]

The relationship with the newborn infant is not one way, for the baby not only returns the mother's intimate inspection but initiates the gaze. And the baby does more than gaze. During the 1990s in the Albert Szent-Gyorgi Medical University in Szeged in Southern Hungary, Emese Nagy produced dramatic evidence of meaningful interactions between adults and infants shortly after birth.[37] In a remarkable videotape[38] she demonstrated interchanges of signals between herself and babies ranging from 3.5 hours to 40 hours old. In some shots Emese sticks her tongue out and the babies stick their tongues out in return. In another part of the tape she raises a finger and the babies respond by raising a finger in return. She lifts her head back and the infants reply similarly. Most striking of all, she was able to observe babies initiating the signalling, apparently trying to evoke a response from the adult. More recent studies by Lynn Murray[39] at the University of Reading show that quite complicated infant communication begins even earlier, not within hours but within minutes of birth. In summary, the physical and emotional intimacy of relationship both inside and outside the womb is intense and it is immediate. It is very obvious that the biological process of becoming a human being is the extreme opposite of an isolated, abstract affair. It is here, in this most natural of processes, that relationship and relational consciousness is made manifest as the primordial mode of being-in-the-world.

LOOKING AHEAD

In the light of what has gone before, let me now summarise the direction in which my argument is about to go, in the form of three consecutively linked assertions. These statements follow logically from the discovery of relational consciousness. The significance of each point will be developed in sequence over the course of the next three chapters:

- As the primordial basis of spirituality, relational consciousness has been the implicit subject of a longstanding clash among psychologists. Underlying a debate that continues to this day are more covert personal differences in religious and philosophical commitments. In Chapter 7 I show how this is particularly clearly illustrated in the arguments amongst the American founding fathers of the psychology of religion, as they debated the nature and validity of religious experience.

- Contrary to what nineteenth-century sceptics asserted, it now seems that spiritual awareness is rooted in our physiological make-up, as Hardy's biological hypothesis implies. Speculation on a neurophysiological basis for spirituality began in the late nineteenth century, but it is only very recently, with the development of sophisticated methods of brain scanning, that progress has been made in identifying that basis. These findings, along with related developments in genetics, are the subject of Chapter 8.
- Chapter 9 presents a pivotal part of the discussion. The political and cultural history of Europe has generated an individualistic ideology that is *in principle* hostile to relational consciousness. This has led to the closing-off and at times repression of spiritual awareness in a way that is unique to those societies that have come under the influence of the European Enlightenment. I argue that relational consciousness is also the underpinning of ethics, hence the ignoring or repression of spirituality has had severely damaging effects on the social coherence of Western society. The current upsurge of interest in spirituality is both a symptom of the malaise and an opportunity to begin the reconstruction of a humane moral commonwealth that respects our relational consciousness.

Part 3

CONFLICT

Chapter 7

PSYCHOLOGISTS START ARGUING ABOUT SPIRITUALITY

One of the main methodological problems in writing about religion scientifically is to put aside at once the tone of the village atheist and that of the village preacher, as well as their more sophisticated equivalents ...

Clifford Geertz[1]

TROUBLE BREWING

Like many young doctoral students, Edwin Starbuck got a rough ride when he gave his initial talk about how his research was progressing.[2] He was the first person to attempt a truly systematic psychological study of contemporary accounts of religious conversion,[3] a task he had begun when he entered the postgraduate department of the Harvard Divinity School in 1893. By the winter of 1894–5 he felt confident enough to accept a request from the Dean to give an interim report of his findings to the philosophy of religion class. About 60 graduate students turned up. Edwin gave them a straightforward account of some of the consistencies that were appearing in his data. For example, the age of conversion appeared to be closely associated with puberty; the phenomena accompanying conversion were like those attending the breaking of habits; there were 'signs of the dissociation of personality and its recentering not unlike the split-personality experiences'[4] described by psychologists. Then the discussion was thrown open to the class. Forty years later Starbuck still remembered the passions aroused:

> That occasion was a sort of christening ceremony for the babe newly born into the family of academic subjects. Some quite

> hot water was thrown into the baptismal font. The first douse of it came from Edward Bornkamp,[5] who rose, his face white with emotion. His first sentence, fervid with the warmth of deep conviction, was, 'It's all a lie!' Laughter broke out there in that dignified classroom ... Of course the attempted damnation of the infant by the first speaker was because its swaddling clothes were only the filthy rags of earthly psychology, ill-becoming the sacredness of religion.[6]

He could probably have predicted it. He already had troubles elsewhere. To help him with his research he had signed up for some courses in the psychology department. His psychology tutor Hugo Münsterberg was 'meticulously helpful' in class but when Edwin tried to pick his brains about the study of religion,

> ... he was antagonistic and finally explosive. He declared that his problems were those of psychology, while mine belonged to theology, and that they had absolutely nothing to do with each other.

Outsiders were also upset. Thomas Wentworth Higginson, an eminent supporter of black emancipation and a keen Darwinian (thus not an illiberal man)[7] happened to stumble across a copy of the questionnaire Edwin used to interview people about their conversion. Outraged, he wrote a letter of complaint to the head of department warning him that it was a forgery. Or if it was genuine, then he wanted to register a strong protest against this 'moral and spiritual vivisection'.

Fortunately the head of department happened to be William James.[8] At first doubtful that Starbuck would find anything interesting, James became fascinated by the data collected by the young man and started giving him advice, most influentially on the general philosophical perspective of his investigation. We can see this in Starbuck's book about his research, where he offered a speculative account of the physical events occurring in the brain in association with the experience of conversion.[9] Clearly he was depending on James' advocacy of the doctrine of 'psychophysical parallelism', that is to say, the idea that every mental event is paralleled by a physiological event in the nervous system.[10] In line with this view, James wrote a preface for Starbuck's book on his research, and remarked,

> Rightly interpreted, the whole tendency of Dr Starbuck's patient labour is to bring compromise and conciliation into the longstanding feud of Science and Religion. Your 'evangelical' extremist will have it that conversion is an absolutely supernatural event, with nothing cognate to it in ordinary psychology. Your 'scientist' sectary, on the other hand, sees nothing in it but hysterics and emotionalism, and absolutely pernicious pathological disturbance. For Dr Starbuck it is not necessarily either of these things.[11]

THE RISE OF EMPIRICAL APPROACHES TO RELIGIOUS EXPERIENCE

'Evangelical extremists' and 'scientist sectaries' are still in plentiful evidence at the present day, but the origins of this dogged divide are longstanding and can be traced back at least to the aftermath of the Reformation in Europe.

Within the Protestant branch of Christianity, particularly in its Calvinist form, there emerged a strong emphasis on the doctrine of predestination. John Calvin taught that God has determined in advance the eternal destiny of every human being; whether they will go to heaven or hell. Since this has been decided from all eternity, and God's will is immutable, there is nothing a person can do about it, for example by leading a virtuous life. Furthermore, strictly speaking, no one can be sure whether they are one of the 'elect' or one of the 'reprobate'. Since the faithful were nevertheless anxious to know if they belonged to the saved, pastoral necessity led to an adaptation suggesting that a personal experience of religious conversion was good evidence of salvation. This motivated the production of careful psychological descriptions of the nature of the process of religious conversion, for the guidance of worried believers and their pastors.[12]

One interpretation of this interest in psychology relates it to a more general shift towards reliance on empirical evidence as the means for obtaining dependable truth, already under way as part of the Enlightenment.[13] By the eighteenth century the primacy of sense experience as the basis of genuine knowledge had become widely accepted in the English-speaking world. In the Eastern United States towards the end of the nineteenth century this axiom was brought to bear on the question of the credibility of religious conversion, as in the case of Starbuck's work. The publication of Darwin's masterwork *The Origin of Species* in 1859 had made the plausibility of religion a hot topic throughout the Western world. But because of the Puritan history of New England the terms of the debate were somewhat different from the parallel disputation in Europe. The centrality of religious conversion as a 'rite of passage' in the Calvinist tradition made the scientific study of religious experience of more than academic interest.

As a consequence, a small group of American students of the psychology of conversion appeared during the last part of the Victorian era. What united these empiricists of religious experience was their belief that psychological science had a positive contribution to make to the study of religion. What divided them was their differing estimates of the validity of religious interpretations of reality, hence their varying emphases on what Paul Ricoeur refers to as the dimensions of *explanation* (sometimes implying a sceptical 'explaining away'), and *understanding* the phenomenon from within a faith context.[14] Apart from James and Starbuck, the group included pre-eminently G. Stanley Hall and James Leuba, but William James is without doubt the towering figure, because of his Edinburgh Gifford Lectures.[15] *The Varieties of Religious Experience* continues to orient our contemporary understanding of spirituality, even though frequently criticised for its individualism and its failure to generate testable scientific hypotheses.[16]

When James was thinking and writing about religious experience he was continuing an investigation which had been begun in New England 150 years earlier by the Calvinist minister Jonathan Edwards. Edwards had been primarily concerned to create a philosophically credible defence of religious feeling using the intellectual tools supplied by the Enlightenment. In particular he developed an adaptation of the sensationalist philosophy of John Locke through postulating a 'sense of the heart'. This is discussed in the first two sections of Edwards' most famous book *The Religious Affections,* published in 1746.[17] The totally novel quality of religious experience led him to dismiss an account of it based on the normal five senses and, as a somewhat aberrant follower of Locke, he proposed the existence of a supra-physical sense given to the believer by divine grace. Edwards held that:

> There is some new sensation or perception of the mind, which is of entirely a new sort, and which could be produced by no exalting, varying or compounding of that kind of perceptions or sensations which the mind had before ... [18]

Edwards' biographer Perry Miller sees him as:

> ... the first and most radical, even though the most tragically misunderstood, of American empiricists. He adopted the sensational psychology with a consistency that outdoes the modern 'behaviorist'.[19]

In a way this is true, but Miller overdoes the point. Edwards' notion of the 'sense of the heart', though analogous to a natural sense, was entirely supernatural, a gratuitous gift of divine grace. His New England isolation and the strength of his prior religious convictions shielded him from the religious scepticism that was more and more to accompany the rationalism of Locke on the other side of the Atlantic. Nevertheless, by the very choice of the term 'sense' Edwards had opened the door to a psychological account of conversion, as we can see when we turn to William James.

James also wanted to produce a defence of religion, though his relaxed attitude to creed would certainly have scandalised Edwards. In a letter to a friend James outlined the underlying purposes of the Edinburgh lectures:

> ... first, to defend (against all the prejudices of my 'class') experience against philosophy as being the real backbone of the world's religious life – I mean prayer, guidance and all that sort of thing immediately and privately felt, as against high and noble and general views of our destiny and the world's meaning; and second, to make the reader or hearer to believe, what I myself invincibly do believe, that, although all the special manifestations of religion may have been absurd (I mean its creeds and theories) yet the life of it as a whole is mankind's most important function.[20]

He quotes from or refers to Edwards, sometimes extensively, on sixteen occasions in the *Varieties*. From the perspective of James (like Edwards, an admirer of 'the good Locke'), the difference 150 years makes is a change in interpretation rather than an absolute change in the substance of the experience. Where Edwards speaks of a 'new sensation', James takes that as a straightforward psychological phenomenon and then makes reference to an eruption of 'subliminal' or 'subconscious' life to account for the novelty and strangeness of the experience. He is drawing not upon Freud, who had yet to make his mark in the United States, but on the speculations of the Englishman F. W. H. Myers on the nature of psychic phenomena.[21] This dependence on Myers is clear in James' celebrated hypothesis about the nature of religious experience, presented in his final Gifford lecture:

> Let me then propose as an hypothesis, that whatever it may be on its *farther* side, the 'more' with which in religious experience we feel ourselves connected is on its *hither* side the subconscious continuation of our conscious life.[22]

And as to 'truth', James the pragmatist held that:

> ... we have in *the fact that the conscious person is continuous with a wider self through which saving experiences come*, a positive content of religious experience which, it seems to me, *is literally and objectively true as far as it goes.*[23]

In his insistence on the literal and objective truth of religious experience, James is intellectually closer to Jonathan Edwards than to modern, purely naturalistic accounts. James' reference to 'whatever it may be on its *farther* side' leaves space for the supernatural. Nevertheless that final get-out clause, 'as far as it goes', sets him apart from Edwards the Calvinist and man of faith. Furthermore, James' location of religious experience within a 'wider self' leaves open the interpretation that such experience is, so to speak, the creation of the subconscious mind and not, as Edwards fervently believed, an encounter with the transcendent God.

The steadily increasing emphasis on psychology rather than theology had the effect of turning attention away from interpretation of experience by reference to a specific historical narrative (in this case the Christian story) and looking towards explanation in terms of 'universals'. As a scientist, James was interested in general laws, those aspects of psychological experience that are, by virtue of a common physiology, applicable to all humanity. This led him to assume rather too easily that religious phenomena from any culture and any part of the world could be subsumed under the term 'religious experience'. Hence the *Varieties* strays far beyond the culturally limited area of Protestant Christian conversion to which Starbuck gave his attention. In a critical article, Cushing Strout catalogues just how far James spreads his net to include accounts of the experience of:

> saints, philosophers, artists and ordinary people, from Protestants and Catholics, Jews, Buddhists, Christian Scientists, Transcendentalists, Quakers, Mormons, Mohammedans, Melanesian cannibals, drug takers, atheists and neurotics, including himself in the guise of an anonymous Frenchman.[24]

The desire to generalise made James an early and important advocate of the 'common core' theory. That is to say he was also implying that spirituality is a primary phenomenon, from which the religions are derived at a second and dependent level. Hence from this perspective spiritual awareness is a more fundamental or basic feature of human beings than their religions. A century after James, we can also see that from this judgement derives the contemporary tendency to see spirituality as completely independent from religion. For a significant number of people that spirituality is 'purer' and less contaminated by the brutalities of history than religion.

A CASE STUDY IN CONFLICT: STARBUCK AND LEUBA

Personal commitments affect the judgements of all scientists in every sphere.[25] This is especially so where matters of intense personal importance are at stake, as is the case with political and religious beliefs. At the overt level, the American empiricists of religious experience appeared to be agreed on the functional utility of religion both to the individual and to society, whatever its ultimate truth. Behind the scenes, the context of their research work was more ambiguous and riven with the complexities that are true of every life history. The conflicting views of Edwin Starbuck and his contemporary James Leuba vividly illustrate the problems of personal bias that still dog modern investigations and to which we shall have to return in the next chapter. They were refreshingly frank about their personal beliefs and made no claim to be neutral and detached about religion in the way that some contemporary scientific students of religion imply of themselves. I therefore want to look in some detail at Starbuck and Leuba's contrasting personalities and their scientific methodologies.

These men knew each other personally. Both were for a time students of G. Stanley Hall, founder of the *American Journal of Psychology* in 1887 and first president of Clark University in Worcester, Massachusetts. Both also knew William James, though Leuba's estimate of James' study of religious experience was much more critical than that of Starbuck.[26]

EDWIN DILLER STARBUCK[27]

Starbuck was brought up as a devout Quaker in the Mid-West, though as a young teacher under the influence of an evangelical revival he recalled going through the forms of a public conversion experienced in the 'Puritan mode'.[28] At the same time his older brother, Edward, 'fell for the same stunts, then went into seclusion for days afterward, feeling that he had quite made a fool of himself'.[29] In the same autobiographical essay Starbuck nevertheless revealed himself as religiously committed, devoted to his Quaker origins, and having throughout his life been aware of 'an interfusing presence'. He did not doubt the potential religious value of conversion (though not necessarily of the sudden kind favoured by Puritans and Pietists) and his researches had an educational as well as a scientific purpose. Though devout, his intellectual perspectives were not confined by his religion. For example, he was familiar with and found congenial Max-Müller's *Introduction to the Science of Religion*.[30]

More than either Hall or Leuba, or any of the lesser students of religious experience at the turn of the century, Starbuck attempted to take the phenomenon seriously and to use empirical methods in its investigation. This meant that he turned away from a content analysis of past religious literature in favour of questioning his contemporaries. He explained that he intended to use an inductive method, that is to say, to gather facts before making generalisations. Explicitly modelling his approach on that of Charles Darwin when he was collecting scientific information from colleagues, Starbuck devised printed questionnaires designed to 'call out' facts of experience rather than opinions. This was 'on the ground that the interpretation of actual experiences would bring us nearer the operation of life-forces than a study of massed opinions'. He was interested in direct personal descriptions rather than doctrinal statements expressed in religious language. By accumulating cases he hoped to 'see into the laws and processes at work in the spiritual life, which do not necessarily harmonise closely with the laws of physics and chemistry, but which nevertheless may have an order of their own'.

Starbuck's book is in two parts, the first being a study of conversion and the second an examination of 'lines of growth not involving conversion'. In treating his data, Starbuck undertook what would nowadays be described as a coding procedure, and where the responses seemed particularly ambiguous he employed a second person to code the material independently. Having produced his coded categories, he presented a large number of cross-tabulations which, had his sampling been truly randomised, would have permitted statistical analysis. He was aware of the limitations of his data, concentrating on results that were clearly apparent on inspection, and noting that they applied only to Protestant North Americans.[31]

The modal age for conversion was sixteen years, with boys being somewhat later than girls. Starbuck's interpretations were two-fold, both psychological and physiological. Teenagers, he said, are impressionable because they lack prolonged experience of life, but also for the first time have the intellectual maturity to grasp the conversion process. Girls are more intellectually precocious than boys; they also mature sexually and begin their adolescent growth spurt before boys and Starbuck noted that this is related to the timing of conversion. Conversion and puberty, he felt, are related in a 'supplemental' manner; that is, conversion seems to come just before or just after puberty, rather than coinciding with it, and he concluded that in some way they mutually condition each other.

Differences between the sexes are also apparent in the motivations leading to the desire for conversion, and in the experiences immediately preceding conversion. Fear of death or hell and other 'self-regarding' motives are more prominent than hope of heaven or love of Christ in leading people to seek conversion. But what would nowadays be called the 'locus of control'[32] differs between men and women. Women are much more likely to report external social pressure as a factor, whereas men tend to refer to an inner conviction of sin and the like. Six times as many women as men reported that they were converted at a church service or prayer meeting, whilst twice as many men as women were converted at home and generally when they were on their own.

Apart from sexual differences, Starbuck thought he detected a difference in temperament between people who do report a conversion experience and those who do not. Quoting from a piece of research conducted by George Coe in New York and published the following year,[33] he cited evidence that people experiencing conversion were more likely than others to be emotional, to exhibit mental and motor automatisms (dreams, hallucinations, uncontrollable laughter, etc.) when at a revival meeting, and to be passive rather than active in personality.

The moment of conversion appeared typically to have an emotionally pivotal quality, though high emotion was not always in evidence. Starbuck quotes one man as saying, 'There was no emotion. It was a calm acceptance of the power of Christ to save.'[34] In most cases, however, strong feelings were evident. The attention seemed to be narrowed and fixed; the senses were more acute ('one person remembers the exact appearance of a pane of glass on which his eyes were resting at the time of conversion').[35] The shift in feeling was typically from a negative mood of 'dejection', 'incompleteness' or 'sense of sin' to 'joy, lightness of heart, peace, etc.'

Significantly, Starbuck noted that although the moment of conversion was traditionally depicted in allegorical terms as a battle between the powers of darkness and the powers of light, with the individual as a kind of onlooker, his respondents never described the experience that way. Nearly all descriptions interpreted it as an internal psychological event, though still couched in conventional theological language. Most people were very inexplicit about the detailed content of their conversion and Starbuck interpreted this as an indication that the process of change was taking place in the 'subconscious'. The two essential factors he abstracted were the need for self-surrender to God, which was followed by a spontaneous outburst of some kind of spiritual illumination.

Starbuck attempted a neurological interpretation of the phenomena so far described. The problem was to explain both the suddenness and the difficulty of making verbally explicit what had happened. Inexplicitness, or perhaps what James called 'ineffability', Starbuck associated with the idea of 'unconscious cerebration', relating it to W. B. Carpenter's account in his book *Mental Physiology.*[36] Carpenter had attempted to explain the familiar phenomenon of spontaneous problem-solving when, after prolonged striving, we cease to attend consciously to the problem and the solution pops up unbidden at a later date. What may be happening, he suggests, following a hint in Carpenter's writing, is that new nerve connections are growing 'in the direction of the expenditure of effort':

> The unaccomplished volition is doubtless an indication that new nerve connections are budding, that a new channel of mental activity is being opened; and, in turn, the act of centering force in the given direction may, through increased circulation and heightened nutrition at the point itself, directly contribute to the formation of those nerve connections.[37]

The down-to-earth physicality of Starbuck's conjecture is striking. As a committed religious believer, he is not embarrassed by embodiment but sees it rather as supporting his cause. Thus he goes on to speculate that, in the person seeking conversion, the striving may be having effects in unconscious ways and, at a certain moment, what has been ripening subliminally bursts into consciousness at the point when the 'new circuit' is completed. Unhappiness and discontent, symptomatic of incomplete links between brain areas, are replaced by a new unification and joy, the conscious aspects of the final readjustment of cerebral innervation around a new organising centre in the cortex.

Turning finally to the repeated assertion by his informants of the necessity of 'self-surrender', Starbuck felt that this was close to the mystery of conversion, and related to the imperfection of the consciously striving self. No doubt the struggle is in the right general direction and helps to 'carry the life' into the anatomical vicinity of the new organising centre. But there is a discord that is resolved when the person relaxes, and lets the nervous energy that has been pent up 'seek its natural and normal channels'. Here, thought Starbuck, the individual must fall back on the larger 'power that makes for righteousness'.

Starbuck continued to use neuropsychological terminology in his references to post-conversion events. Almost universally reported was an increase in feelings of altruistic love for God and man. What had previously been 'reflex' responses to religious and moral precepts became 'real' and personally appropriated. It was as if what had existed as 'reflexes in the lower nerve centres' had been taken up as factors in 'higher cerebral activity'. This change, Starbuck observed, took place at the point where the young person is about to enter adult society and where 'physiological awakening has announced the possibility of parenthood and citizenship'.

The appearance of altruism in connection with religious conversion was thus viewed by Starbuck as, however inexplicable in physiological detail at the time he was writing, a perfectly normal and natural process that had evolutionary survival value.

> It is natural [from the perspective of evolutionary survival] that in adolescence there should be a rapid development, which either furnishes some of the elements that directly enter into religion, or brings the individual suddenly into such ripeness of mental capacity that religious impulses may have an adequate organ for their reception and expression.[38]

In spite of his lengthy and detailed study of conversion, Starbuck had a degree of scepticism about the necessity of sudden conversion as worked upon by revivalists, compared with a more gradual growth into religious consciousness. He was suspicious of the manipulation of feelings, combined with the pressure to conform, something of which Jonathan Edwards was not unaware 150 years earlier. Revivalists use methods uncomfortably like those applied by the crowd leaders described in Gustave Le Bon's study of mass hysteria, said Starbuck.[39] They employ *affirmation* of statements without rational support, *repetition* of statements, and depend upon the *contagion of feeling* within the crowd to spread their effects. Neurologically, something different from true conversion may be happening, that is, 'a relaxed state in the cerebral centres which frees the lower centres from inhibitory control by higher, and thus renders the mind more suggestible'. When this happens to the individual, he becomes susceptible to hypnosis by the evangelist. Here Starbuck is using the kind of discernment one might expect of a spiritual director to identify inauthenticity.

Nevertheless, he was not as hostile to the religious validity of sudden conversion as Boris Sidis who, in reviewing Starbuck's writing said,

> What Mr. Starbuck does not realise is the fact that it is not healthy, normal life that one studies in sudden religious conversions, but the phenomena of revival insanity.[40]

In his autobiography Starbuck noted that a similar criticism appeared in the *Philadelphia Medical Journal.* But he felt he had the balanced view, believing that he was examining an anthropologically normal phenomenon, distorted at times by revivalism. On the other hand, religionists were too inclined to see everything within religion as healthy. Starbuck did not attempt to impose these discriminations in his professional writing, but let people's own reported view be the criterion of what was happening at the level of consciousness. Thus, in important respects he resembles a modern phenomenological psychologist in his methodology, though he was also completely overt in his affirmation of the religious perspective. The ultimate test of the validity of religious experience, he felt, was whether in the long run, in the individual and in groups of individuals, it contributed to permanent growth.

JAMES H. LEUBA

Leuba had a much more sceptical perspective on religion than Starbuck. It may be significant that he was not native to America, being born in Neuchâtel in Switzerland, where he spent his childhood and youth before emigrating to the United States. After early difficulties, he became a student of Stanley Hall at Clark University, and on completing a PhD and further psychological researches he obtained an appointment at Bryn Mawr College in Pennsylvania where he remained as professor of psychology for the rest of his working life.[41]

His origins were fairly humble. His father was a watchmaker. Although both his parents were church members and serious about religion, Leuba's father had a curiosity about life and a critical attitude towards religious ideas that in Leuba's opinion would 'have made of him a downright heretic had he possessed enough knowledge'. Leuba recalled how, in his youth in Neuchâtel, he was prepared for first Communion through the teaching of a 'mitigated Calvinism'. Without commitment to these beliefs, he was told, there was no salvation. Perhaps he had learned from his father's independence of mind, for he was puzzled and miserable at his failure to comprehend the 'fantastic doctrines' that he was required to affirm. As an inexperienced boy being taught by a renowned New Testament scholar who was also a member of a prominent local family, he could not find enough inner resolution to say to his teacher that he did not believe what he was being taught. In his youthful confusion he could not decide whether his misery was due to a defect in himself or in the religion he was taught.

Some time later Leuba came across the Salvation Army in Neuchâtel. A group of Salvationists had been sent to the town on an evangelical mission and the supposedly pious citizens received them with great intolerance. They were jeered at and punched, had stones thrown at them and finally were thrown into prison, but remained undeterred. Leuba was drawn to them by their sincerity and courage and in the end underwent a kind of conversion, which he described as being more moral than religious.

> The moment came when I could no longer resist the appeal of the moral ideal they were for ever holding up before us: no compromise with evil, no half-way measure, no divided self; every impurity had to be disavowed, the divine Will alone was to rule. In the conversion through which I passed, the doctrinal background, presented so vividly and tirelessly by my friends of the Army, played a remarkably small role. It is the moral ideal itself which moved me. I saw it as an Absolute which it was my duty and privilege to realise. There was, in addition, an acute sense of guilt for having fallen short of a perfection regarded as attainable. This ethico-religious experience was perhaps the most beneficial one of my life; it was certainly the most violent one.[42]

Following this experience, Leuba became engrossed in studies for a BSc degree and, developing more and more admiration for the scientific method and Darwinian ideas, imperceptibly drew away from his earlier religious adherence. After he migrated to the United States he was still sufficiently in touch with religion to act as General Secretary to the French YMCA in New York, but was relieved to be able to resign this post after two years. When he saw the opportunity to enter Clark University in Massachusetts, he felt it was the chance he had been waiting for:

> The physical and the biological sciences were throwing a wonderful light on the mysteries of the world, but what I wanted to understand above all was the working of the human mind. Psychology, thought I, would in particular clear up, among other things, the wonders of conversion and of the religious life in general.[43]

Not surprisingly, when the opportunity arose to select a topic for a doctoral dissertation, Leuba chose 'conversion'. What seems to have happened during the evolution of Leuba's mature views was an acceptance of the Comtean interpretation of religion. That is, he saw it as something which people outgrow as they come to understand the scientific point of view. To hold this opinion is to adhere to a particular metaphysical stance,[44] as is evident in the style of presentation which Leuba used in his PhD, published in the *American Journal of Psychology*.[45]

His interpretations are apparently based on only thirteen responses to his questionnaire, supplemented by eight more accounts abstracted from the literature. The set of questions is itself much briefer than that of Starbuck, and was sent to persons thought to have 'experienced religion', that is, mission leaders and pastors and also to a number of Protestant religious journals. One may speculate therefore that the number of replies he received was a good deal larger than those he felt able to use in his published report.

On this foundation Leuba then attempted a sequential analysis of the process of sudden conversion, beginning with the sense of sin, which he attributed to the effects of Calvinistic doctrine, and which leads to irrational fears and bodily disorders. According to Leuba, the order of the stages followed closely on that described by both St Augustine and Jonathan Edwards, consisting of self-surrender, faith, justification, joy and the appearance of newness. The last part of the paper examined the role of the will, the doctrine of faith, determinism and the doctrine of divine grace. Though these topics appear in the work of Starbuck, in Leuba's treatment there is, by comparison, much more theology and philosophy, and less interest in the empirical data. The thesis Leuba was proposing is nevertheless clearly stated. Like Starbuck, he wished to be seen as a naturalist, but he sought to show that the phenomena associated with religious conversion could be accounted for exhaustively within the parameters of natural science. He thus differed from Starbuck in being concerned with 'explanation' in scientific terms rather than in religious or theological 'understanding'.

He was interested in homologies between religious conversion and sudden conversion from drunkenness without benefit of religious belief or experience. However, the briefness of his treatment meant that he skirted round the possibility of differences in the detailed phenomenology of the cases he proposed as comparable. He also avoided entering into the kind of physiological speculation that Starbuck used when, for instance, the latter attempted to discriminate between conversion and mass persuasion. In his desire to explain religious phenomena naturalistically, Leuba became less and less concerned with the empirical investigation of contemporary phenomena, in favour of an historical study of the writings of the mystics. There is thus some sense of searching for and fitting data to a preconceived theoretical structure. Though this is hardly absent from Starbuck's work, he was much more of a phenomenologist in his attitudes than Leuba.

Leuba was an extremely prolific writer and *The Psychology of Religious Mysticism*,[46] not published until 1925, is his most important and interesting work. It seems likely that he himself saw it as his most substantial academic contribution, since it is primarily a republication of two articles which appeared in the *Revue Philosophique* more than twenty years previously, in 1902, the same year as James' *Varieties* was published. Leuba was especially interested in the great European Catholic mystics because he felt they exhibited the tendencies manifested in the Protestant conversion experience (and hence presumably his own experience) to an extreme degree.

Nevertheless, he wished to present such experience as evolving from more primordial beginnings in what he referred to as the drug-induced hallucinations of 'savage' societies and in the altered states of consciousness produced by physical techniques such as those of yoga. His treatment of these purported beginnings is very perfunctory and error-prone. For example, he refers indiscriminately to Hindus as Buddhists and appears to think the *Yoga Aphorisms* of Patanjali is a Buddhist work. But the line of argument he develops is clearly biological in its metaphors. The ecstasies of the saints are not a peculiarity of Catholicism or even Christianity, but a potential in the species. 'Mystical ecstasy', he says at one point, was not an invention of the Western mystics, but, 'the joint produce of chance discoveries and empirical gropings begun already in man's infancy'.[47]

The Christian mystics on whom he spends most time are Henry Suso, Catherine of Genoa, Mme Guyon, Teresa of Avila and Margaret Mary Alacoque, with most space given to the Quietist Mme Guyon. 'What do these mystics really want when they say they want God?' he asks. The answers are: self-affirmation; the need to devote one's self to somebody; the need for affection; the need for peace and unity; but most of all, organic sexual needs. Because of an ideal of chastity implanted from an early age, the mystics were prone to ignore or deny their sexuality. In addition, biographical data reveal in some cases, for example those of Catherine of Genoa and Mme Guyon, that although married, the sexual side of their marriages was unsatisfactory.

Many of the mystics' descriptions of their religious ecstasies utilise the language of eroticism and Leuba says that this is indeed accurate, for the sexual organs are directly involved:

> A sufficient acquaintance with facts that come to the knowledge of diligent students of the sexual life, establishes the conviction that the sexual organs respond in some degree to all tender thoughts, with a certainty and delicacy which will appear impossible to those not well informed.[48]

Because they were not aware of the bodily origin of the sensuous enjoyment, said Leuba, the mystics 'gave themselves up to it with great relish and complete abandon'. The best explanation of their behaviour, he says, is that it is autoerotic and they are suffering from 'erotomania'. Furthermore:

> ... they also had to live a life of continence and they were for years divided souls with regard to the things which most matter to man. Because of the ideal they had formed and of the method of life they had chosen, their deepest instincts and desires could not be gratified in the ordinary way. And they aggravated repressions and conflicts, in themselves sufficient to cause a variety of psychoses, by excessive and persistent ascetic practices, and thus exhausted themselves to the point of inanition.[49]

For the remainder of the book, Leuba draws together evidence in support of the view that mystical experience, so-called, is either to be associated with a pathological mental condition or is due to misattribution. Many of his ideas resemble quite closely those of modern attribution theorists.[50] Thus, he notes that spontaneous ecstasies of the type reported by religious people are very similar in content to an epileptic 'aura' just before a fit. The aura is caused by a physiological disorder. It occurs suddenly and unexpectedly to a passive subject. It is often experienced as bringing illumination, revelation or *noesis* and it is ineffable. Such characteristics might suggest superhuman causation, but that is not the ascription of the medical profession. It is interpreted as a symptom of disease. St Paul's raptures could probably be explained in the same way, adds Leuba.

Sometimes the Jamesian 'will to believe' may lead to more or less conscious self-deception. Leuba quotes a 'pathetic chapter in the life of a lone woman', who had lost her religious belief and felt abandoned. She deliberately set to work to re-acquire the sense of God's presence which she had not had for twenty years. The strategy adopted was to use the sceptic's prayer, 'O God, if there is a God, save my soul, if I have a soul.' After a week's continuous use of this, the old sense of God's presence came over her with great power. Leuba adds that she cared less and less whether the sense of presence was 'subjective or not'.

One piece of empirical research that Leuba did attempt with his students at Bryn Mawr College concerned this 'sense of presence'. Students were blindfolded and seated in a thickly carpeted room and asked to note whether or not someone had entered the room. At times they were convinced of the presence of someone else, even when there was no one there. Leuba explained this as resulting from the subjective physiological impressions one has when one is *expecting* someone else to be present. In other cases, he says, it is simply due to a 'brain-storm'.

However humble, or perhaps pathological, such a sense of presence may be, Leuba went on to claim that it had remarkable effects on the lives of religious people:

> The development of the mystical technique for the realisation of a quasi-physical presence of the Perfect One constitutes the most remarkable achievement of religion in man's struggle to overcome adverse external circumstances, his own imperfec-

> tions, and those of his fellow men. It is one of the outstanding expressions of the creative power working in humanity. It is paralleled in the realm of reason by the development of science. Both lead, if in different ways, to the physical and spiritual realisation of man.[51]

In the light of the curious mixture of hardheadedness and negligence of the phenomenology of religion displayed by Leuba, I find it difficult to believe that the above quotation is entirely ingenuous. He died in 1946, a year before Starbuck, and his obituary in the *American Journal of Psychology* notes his running battle with the religious establishment. Over the years this added an increasing note of polemical reductionism to his comments on religion. I am therefore inclined to interpret the statement as a somewhat forced attempt to appear even-handed.

CONCLUSION

I have gone into the background and personal opinions of Starbuck and Leuba in considerable detail because their stories demonstrate very clearly the opposite ends of a spectrum of belief. It is a spectrum that is still crucially important when it comes to making judgements about the validity of any piece of research in the field of spirituality. In assessing the contributions of Starbuck and Leuba it is apparent that Starbuck was more competent and careful as a research worker than Leuba. Though Leuba was much the more productive of the two men, at least in publications in mainstream psychology of religion, he was opinionated, careless about detailed accuracy and inclined to ignore data that was unhelpful to his case. Yet, during the early part of the twentieth century, he held an overriding position in the psychology of religion. His status meant that although his researches turned away more and more from directly empirical investigations, he had a dominant role in the formation of attitudes within his field. Pierre Janet considered him, along with Durkheim, its greatest living exponent.[52]

Personal interest both plagued and motivated the central protagonists in this argument. In contrast to Leuba, Starbuck was engaged upon the reconciliation of science and religion. His methodology suggests that he may have dimly recognised (to quote from Paul Ricoeur) that 'explanation alone can be reductive, but that understanding alone remains vulnerable to uncritical individual or corporate illusion or self-deception'.[53] Ricoeur's hermeneutics of suspicion require the acceptance of critical (including psychological) explanations of religious phenomena, combined with an understanding that may operate at a post-critical level, to achieve a sophisticated 'new naivety'. But it seems that the empirically meticulous Starbuck was to a degree stuck in a pre-critical credulity which laid him open to the charge of failing to take full account of the Enlightenment critique of religion.

Even with all his carelessness and his methodological limitations, Leuba's explanations were more in tune with the times. The force of his opinions, along with the advent of psychoanalysis and behaviorism during the second and third decades of the twentieth century meant that the poles of 'explanation' and 'understanding' in the psychology of religion were virtually torn apart by 1930. In one of Leuba's favoured mouthpieces, the *Psychological Bulletin*, the last annual review of the field of the psychology of religion appeared in 1933.

It was not until the 1960s that the same degree of energy began to return to the empirical investigation of spirituality. Since then the field has been transformed, particularly through developments in neurophysiology and genetics, as we shall see in the next chapter. In spite of these advances the same problems of personal bias continue to haunt the field – transparency as to personal stance is not always available to us and sometimes investigators give the impression that they are ideologically neutral, in spite of our modern understanding of the impossibility of detachment. It will be important whenever possible to become aware of the background of scientific and religious beliefs of the researchers under consideration when making judgements on their findings.

Chapter 8

MODERN SCIENTISTS WIDEN THE ARGUMENT

Who are you?

Question from a personal boundary setting exercise

CHEMICAL ECSTASY

Edwin Starbuck got himself in trouble with his fellow theologian, Edward Bornkamp, when he suggested that there was a connection between spirituality and the physiology of the brain. Oddly enough, many atheists would have sided with Bornkamp. Ludwig Feuerbach had made the point explicitly during a lecture series in Heidelberg in 1848.[1] Turning a critical eye on the view that spirituality is innate, he scornfully denied the possibility 'that man has a special organ of religion, a specific religious feeling':

> We should be more justified in assuming the existence of a specific organ of superstition. Religion, that is, the belief in gods, in spirits, in so-called higher invisible beings who rule over man, has been said to be as innate in man as his other senses. Translated into the language of honesty and reason, this would only mean that … superstition is innate in man. But the source and strength of superstition are the power of ignorance and stupidity.[2]

A little over a century after Feuerbach's lectures, during the 1960s, fashionably dishevelled hippies began proclaiming the dawning of the Age of Aquarius, urging people to turn on to transcendence by ingesting the right kind of drug, often lysergic acid or LSD. Whatever we now think of their advice, their advocacy of chemical ecstasy gave a hint that, in spite of Feuerbach, spiritual experience might have something to do with the biochemistry of the brain.

Alerted to the extraordinary effects of psychedelic drugs in producing altered states of consciousness, many people experimented with them, including some psychologists.[3] One of the most notorious was Timothy Leary, for a short time a lecturer in the Psychology Department at Harvard University. Leary had a reputation as a wild man and was thrown out of the department in 1963 after complaints from parents about his distribution of hallucinogens to his students. Amidst his excesses Leary supervised what is probably the most interesting of all studies on the relation between psychedelic drugs and spirituality, the doctoral research of a young theology student called Walter Pahnke.[4]

At the core of Pahnke's thesis was an elaborate experiment, popularly remembered as the 'Miracle of Marsh Chapel', conducted during Holy Week in the chapel of Boston University. Pahnke selected a group of 20 students from Andover Newton Theological School, near Boston and, using a double-blind procedure, gave half of them a prescribed amount of the hallucinogenic drug psilocybin and the other half a placebo. They then took part in Good Friday devotions in the chapel and afterwards were asked to complete a lengthy questionnaire to determine what they had experienced during the service. Broadly speaking, Pahnke reported that those who had been given the placebo experienced nothing out of the ordinary, whilst a significant number of those who had ingested psilocybin felt their religious understanding had been dramatically deepened.

Pahnke was awarded a doctorate for his thesis by Harvard University in 1963 and in 1991 Rick Doblin published a long-term follow-up and critique.[5] Doblin was fortunate enough to be able to make contact with 19 of the original 20 members of the group, many of whom were by that time well into their careers as ordained clergy. Doblin had some serious criticisms, including the fact that Pahnke had concealed an unfortunate incident that took place during the experiment. One of the psilocybin takers had become highly agitated, to the extent that Pahnke felt he had to give him a tranquillising injection to calm him down. This man was also the one member of the group who declined to be interviewed by Doblin. Nevertheless, approximately a quarter of a century after the meeting in Marsh Chapel, Doblin found that with that one exception,

> The experimental subjects unanimously described their Good Friday psilocybin experience as having had elements of a genuinely mystical nature and characterized it as one of the high points of their spiritual life …

On the other hand,

> Most of the control subjects could barely remember even a few details of the service.[6]

Several of the psilocybin takers commented on how their experience of 'unity with all reality' strengthened their concern for social justice and care for the environment. The feeling of timelessness also reduced their fear of death and gave them courage to enter into the political struggle. Doblin concluded:

> The long-term follow-up interviews cast considerable doubt on the assertion that mystical experiences catalysed by drugs are in any way inferior to non-drug mystical experiences in both their immediate content and their long-term positive effects.[7]

It is natural to feel suspicious about what seems like a mechanical procedure, swallowing a drug to 'switch on' an experience of transcendence,[8] and Pahnke's experiment was daring. On the other hand it was quite remote from the casual practice of ingesting drugs for kicks or as a means of escape. The research seems on the whole to have been conducted responsibly, with the possible exception of the care of the individual who became unduly agitated. The participants volunteered freely and furthermore the context of a religious service placed them in a cultural milieu that cohered with both their core beliefs and the metaphors of transcendence with which they were familiar. The descriptions of their experiences under the influence of psilocybin are recognisably like the instances of relational consciousness I have discussed in earlier chapters. Furthermore, if measured by long-term outcomes on the biblical criterion 'By their fruits shall you know them', they probably pass the test.

Nevertheless, for those unlikely allies, sceptics and religious fundamentalists, the demonstration of a biological process underlying a supposed spiritual experience allows them to dismiss it as 'nothing but' physiology. From my perspective, both parties are reductionists. They hold to their views at the expense of denying important dimensions of what it is to be an embodied human being. To me as a zoologist it would be very surprising if spiritual experience were not mediated via our physiology. Two recent scientific developments have allowed scientists to explore this issue more deeply: the mapping of the human genome and the invention of brain scanning techniques.

THE HUMAN GENOME

TWIN STUDIES

One way of getting solid evidence on whether a human trait is inherited or merely acquired from the environment is to study twins. The method depends on a crucial difference between identical and non-identical twins. Identical or 'monozygotic' twins originate from the splitting apart of a single fertilised egg and have inherited exactly the same set of genes, or 'genotype'. Therefore any differences between them must be due to environmental factors. Non-identical or 'dizygotic' twins come from two separate eggs, each fertilised by a different sperm and consequently there is no reason to expect any more genetic similarities between them than between ordinary brothers and sisters. Using this information along with sophisticated methods of statistical analysis it is possible to put a figure on the relative importance of genetics and environment in the case of any human characteristic.

There have been several recent attempts to assess the inheritance of religious behaviour (as opposed to spiritual experience) using large samples of twins. In 1999, along with four colleagues, Brian d'Onofrio, a young psychologist at the University of Virginia,[9] reported on a genetic analysis of religious data from the 'Virginia 30,000' (a sample of 14,781 twins and their family members). The dimensions they measured included church attendance and a cluster of attitudes characteristic of the Religious Right in America, concerning (a) the Moral Majority (b) school prayer (c) censorship (d) racial segregation and (e) hostility to pacifism. On the whole, church attendance and the array of attitudes associated with Right Wing religion seemed to be largely determined by the social norms of one's family and friends and not by heredity. Unfortunately this tells us little or nothing about the spiritual life of the members of this very large sample because no distinction was made between formal religious practice and spiritual experience.

Another report appearing in the same academic journal in 1999 took the investigation a little further[10] In a study of seventy-two sets of twins, split into approximately equal numbers of monozygotic and dizygotic pairs, Thomas Bouchard and his colleagues found that intrinsic religion (approximately, religion that is deeply personally meaningful) seemed to have a genetic component, whilst extrinsic religion (practised primarily out of social conformity) did not. Bouchard did not investigate report of spiritual experience in the sample, though we might assume that intrinsic religion is more likely to be associated with spiritual awareness, hence supporting the genetic link.

Much firmer evidence comes from the work of the geneticists Katherine Kirk and Nicholas Martin at the University of Queensland in Brisbane and their colleague Lindon Eaves[11] at the Virginia Institute for Psychiatric and Behavioral Genetics. Kirk and her colleagues sent letters to more than 2,200 pairs of twins[12] inviting them to complete a questionnaire including a measure of 'self-transcendence' devised during the 1980s by Robert Cloninger, a psychiatrist at Washington University Medical School in St Louis. Cloninger saw self-transcendence as having three elements. The first is 'self-forgetfulness', meaning the ability to lose oneself in a task to the point where it seems to 'do itself'. It is as if the boundaries between the self and the rest of the world simply dissolve. A second element, obviously connected with the first, is 'transpersonal identification', meaning a sense of being in a unity with the rest of nature. People with high levels of transpersonal identification feel protective of other people and of the natural world because of their experience that somehow it is part of them. Thirdly, Cloninger identified 'mysticism' as an important element, meaning a strong personal concern with the profoundest mysteries of human existence, often accompanied by an intuitive sense of deep meaningfulness.

When I saw what Cloninger meant by self-transcendence I was fascinated. His descriptions could have been transcribed directly from many accounts of relational consciousness in our archive. Not surprisingly, I was eager to see how the Australian twins scored on the self-transcendence scale. The results were very clear. Identical twins were approximately twice as likely to get similarly high or low scores on self-transcendence as compared with non-identical twins. This is exactly what you would expect if self-transcendence were genetically mediated. Commenting on this result, the leading American geneticist Dean Hamer remarked:

> The implication is that spirituality, at least as measured by self-transcendence, doesn't result from outside influences. Contrary to what many people might believe, children don't learn to be spiritual from their parents, teachers, priests, imams, ministers or rabbis, nor from their culture or society. All of these influences are equally shared by identical and fraternal twins who are raised together, and yet the two sets of twins are strikingly dissimilar in the extent to which they correlate for self-transcendence. In other words, William James was right: Spirituality comes from within. The kernel must be there from the start. It must be part of their genes.[13]

Kirk and her colleagues also checked to see if there was a genetic influence on church attendance. Like the other two groups of researchers, they found that the main influence was the shared environment. The likelihood of one twin going to church if their co-twin went was almost the same whether the twins were identical or fraternal.

Further evidence of the major importance of environment on church attendance is the fact that attendance levels differ very widely between populations that are quite similar in the frequency with which they report spiritual experience. Americans are as much as five times more likely than British people to attend church regularly, yet as we saw in Chapter 2, report of spiritual experience is at least as high in Britain as in the USA, sometimes higher.

THE 'GOD GENE'

If spiritual awareness is inherited, in principle it ought to be possible to track down the part of the human genome concerned. There has been at least one widely reported attempt to do this. The cover of *Time* magazine for 25 October 2004 carried a drawing of the head and shoulders of a woman, her eyes closed in meditation. Perhaps she is Indian, for at the midpoint of her forehead, where devout Hindus paint a *bindi* or sacred symbol, she has an unusual mark in the form of a double helix. The image combines the structure of DNA and a representation of prayer. It was highlighting an article about Dean Hamer, who had recently published a book entitled *The*

God Gene: How Faith is Hardwired into Our Genes.[14] Hamer stirred up a predictable storm of protest from fundamentalist religious groups. Feuerbach might have been amused or disconcerted to know that they were insisting that religious belief has nothing to do with our biology. Critics from mainstream science were not much kinder. Carl Zimmer reviewed the book in the *Scientific American* magazine and concluded by suggesting that it would be better titled 'A Gene That Accounts for Less Than One Percent of the Variance Found in Scores on Psychological Questionnaires Designed to Measure a Factor Called Self-Transcendence, Which Can Signify Everything from Belonging to the Green Party to Believing in ESP, According to One Unpublished, Unreplicated Study.'[15]

Well, yes. But a careful reading of Hamer's book shows that his thesis is not as ridiculous as Zimmer implies. The working out of the total structure of the DNA that goes to make up the human genome is a very recent achievement, so it might seem extremely ambitious to attempt to track down the genetic underpinning of our spiritual life. Nevertheless, that is what Hamer attempted. As Chief of the Section on Gene Structure and Regulation in the Laboratory of Biochemistry of the National Cancer Institute in Bethesda, Maryland he was probably as well equipped as anyone in the world to do it.

Hamer decided to investigate a population of same-sex siblings (brothers or sisters) who had been recruited for studies of smoking behaviour (1,001 subjects) and personality (623 subjects). That is to say, in terms of spiritual awareness they were a relatively random group. As part of a larger questionnaire, they were all given Cloninger's self-transcendence scale to fill in. The next stage was harder. Collecting DNA is not the problem; it is done every day in genetics laboratories all over the world. A swab of saliva treated in the right way (Hamer gives a simple recipe in his book)[16] produces easily visible white, silky threads of the chemical that contributes so powerfully to the shaping of our lives. The difficulty is that there are around 35,000 genes in the human genome and the functions of the majority of them are still unknown. Tracking down possible candidates as the underpinning of self-transcendence was going to require a lot of work as well as some inspired guesswork.

There were two hints to go on. Hamer knew that drugs that produce altered states of consciousness resembling the experience of self-transcendence act on a group of chemicals in the brain called monoamines. Then, with the serendipity that occasionally happens in the laboratory, during a coffee break a colleague started talking about a gene called VMAT2[17] he had been working on, that was involved in the storage of monoamines in the brain. When Hamer and his colleagues then took another look at their research material, concentrating on differences in just one chemical base[18] in the gene, in his own words,

> [they] hit pay dirt. There was a clear association between the VMAT2 polymorphism and self transcendence. Individuals with a C in their DNA – on either one chromosome or both – scored significantly higher than those with an A. The effect was greatest on the overall self-transcendence scale and was also significant for the self-forgetfulness scale. With transpersonal identification and mysticism, the effect was in the same direction but just short of statistical significance.[19]

Whatever else Hamer did, he certainly didn't discover 'the God gene'. Part of the scandal was caused by the provocative title of his book, possibly chosen by the publisher as an eye-catcher to promote sales. What he *did* do was to identify one gene out of a probable gene complex involved in the management of monoamine release in the brain. His work is a first step in identifying the physical substrate that makes relational consciousness possible. As such it has no bearing either positively or negatively on the existence of God, but it is another piece of evidence reducing the plausibility of the arguments of those who dismiss spiritual experience as nothing but fantasy.

NEUROPHYSIOLOGY

If the findings of the genetic research I have discussed are correct, how do these inherited factors translate into the spiritual awareness of the fully developed human being? One promising way forward lies in the new field of study labelled by some of its proponents 'neurotheology'.[20] The term is unfortunately a gift to the popular media, who trivialise the research with headlines like 'the search for the God Spot'. The sensationalism conceals the fact that in the midst of the hype there is a programme of serious empirical study. Amongst the more notable scientists who are contributing to the investigation are V. S. Ramachandran at the University of California in San Diego, Michael Persinger at the Laurentian University in

Sudbury, Ontario, Andrew Newberg in the University of Pennsylvania, Mario Beauregard at the University of Montreal, and Nina Azari at the University of Hawaii.

TEMPORAL LOBE EPILEPSY

One starting point is curiosity about the disease of epilepsy. Students of temporal lobe epilepsy have long known that people with this condition quite often report intense religious experiences in association with a seizure. Working with epileptic patients in San Diego, Vilayanur Ramachandran noticed that this was the case with about a quarter of them.[21] He did an interesting follow-up with the help of a very small group of patients by checking their emotional arousal when shown pictures of a variety of subjects. The way this is done is to measure small changes in the electrical conductivity of the skin, since it is known that conductivity (galvanic skin response or GSR) goes up when people are emotionally aroused, for example when they are shown pictures of sexual or violent behaviour. The people suffering from temporal lobe epilepsy that Ramachandran tested responded with an especially strong GSR to religious symbols, and were less inclined than other people to be aroused by signals representing sex and violence.

Ramachandran offers a number of speculations about the possible reasons for the phenomenon, whilst reminding us that his suggestions are tentative since only a very small group of patients was studied. From a straightforward religious perspective it could be that God is communicating with them particularly vividly. Or perhaps the seizure produces false signals between a part of the brain called the limbic system and the sensory centres in the cerebrum, giving intensity to what in normal brain activity operates at a lower level.[22] Or there could simply be an overflowing of emotion that is interpreted religiously. Reflecting on these possibilities, Ramachandran suggests briefly and tentatively that religious or spiritual awareness has evolved through the process of natural selection, though without elaborating on the idea, and apparently unaware of Hardy's much more detailed evolutionary explanation.

Another person to have taken an interest in the temporal lobe hypothesis is Michael Persinger. Persinger has written a great deal on this subject and is best known in the popular media for his invention of a helmet that he claims can induce religious or spiritual experiences by producing brief microseizures in the temporal region of the brain. The device looks like an ordinary motorcyclist's helmet but is fitted with a set of solenoids through which a magnetic field is projected into the brain of the subject, the appropriate wavelength being adjusted by the use of a computer. In a scientific paper published in 1983[23] Persinger explained his idea:

> According to the hypothesis, the actual mystical or religious experience is evoked by a transient (a few seconds) very focal electrical display within the temporal lobe. Such temporal lobe transients (TLTs) would be analogous to electrical micro-seizures.[24]

He suggests that there is a naturally occurring range of sensitivity to electrical stimulation, with 'temporal lobe sensitives' being those most likely to experience altered states of consciousness. A stream of journalists and scientists, including sceptics like Richard Dawkins and Susan Blackmore, have made their way to what Persinger calls his 'dungeon' to try out the helmet for themselves. He claims that a fair proportion of them, amongst whom Blackmore includes herself, have had extraordinary experiences.[25] Some say they have felt a sense of presence and, rather less often, seen visions that are at least partly determined by their cultural expectations. Dawkins reported that he experienced nothing more than a mild tingling sensation, as befits a devout atheist who had also scored low on Persinger's prior pencil and paper test to assess temporal lobe sensitivity.[26]

Persinger's further conjectures wander into increasingly idiosyncratic speculation. He suggests that one source of naturally-occurring experience of the supposed supernatural is random geophysical movements in the earth, creating electrical fields that impact upon the brain. TLTs, he says, can generate negatively or positively charged emotions and can retrieve infantile memories of parents that provide a source of God images. Persinger's research has a spectacular and eccentric quality that gets him a lot of publicity. In the light of recent findings by a group of neurophysiologists in Philadelphia, it now also seems likely that his concentration on the temporal lobes is marginal to an understanding of relational consciousness.

THE PHYSIOLOGY OF RELATIONAL CONSCIOUSNESS

More directly relevant to an exploration of the physical basis of relational consciousness is the work of Andrew Newberg, who is in charge of the Nuclear Medicine Department in the University of Pennsylvania Medical Center in Philadelphia. In 2001, along with a group of researchers including the late Eugene d'Aquili,[27] he published the results of a study of the changing patterns of brain activity in eight experts in Tibetan Buddhist meditation, all with at least 15 years' experience of regular daily meditative exercises.

Newberg devised a method of detecting the distribution of physiological activity in the brain once his volunteers had entered the meditative state. The device that allowed him to do this was a SPECT (Single Photon Emission Computed Tomography) scanner. This is a machine commonly used in nuclear medicine to detect the levels of metabolic activity in different parts of the body, as monitored by following the movements of a radioactive substance injected into the bloodstream. Newberg's first step was to make a brain scan of each meditator whilst they were in an everyday frame of mind, to give himself a baseline to compare with the changes that take place during meditation. Then came the tricky part; catching the stage when the volunteers entered the meditative state. To give them the best chance of entering deep meditation Newberg set up a quiet room provided with whatever was helpful in the way of pictures, incense or music. After having a catheter line inserted into the left arm to allow radioactive fluid to be injected into a blood vessel at the appropriate time, the volunteer was left alone to settle down and begin meditating.

Meanwhile, Newberg and his colleagues sat in a lab just down the corridor, waiting for the appropriate time to inject the fluid along a tube leading to the catheter. When the meditator entered a deep meditative state they gave a brief signal and at that moment Newberg injected a radioactive substance called Tc-99m-HMPAO into the bloodstream. Tc-99m-HMPAO is one of the commonest radioactive tracers used for brain-SPECT imaging and it has the unusual property that it remains for a period of several hours, attached to the point where it first locked into the brain This allowed the meditators to finish their meditation, then move to the SPECT camera without losing the information about the regional differences in blood flow in the brain. Newberg's team then used a variety of methods to assess the differences in blood flow between baseline and meditation in 15 regions of the brain, each region chosen because there was reason to expect that it might be involved in the meditation process.

To the layperson, the technology employed in modern scanning devices is formidably sophisticated, but their limitations are still great. In addition the changes in blood flow were relatively small, amounting to a few per cent, thus making measurement a tricky matter. Add to that the fact that the activity of the brain is extremely complex and by no means fully researched and it becomes clear that care must be taken in interpreting Newberg's findings. Newberg is appropriately modest about the findings, pointing out firstly the small number of subjects he was able to use, because of the difficulty in identifying highly experienced meditators. Secondly, the study measures regional cerebral blood flow at a single point only, during a lengthy process of meditation.

Nevertheless the Philadelphia team feel confident that they have detected a number of genuine changes in brain activity during that process, suggesting that meditation involves a whole complex of actions. Two of them stand out and are of particular importance for our understanding of relational consciousness. Before meditation started, the SPECT scan showed high activity in the left superior parietal lobe of the cerebral cortex. This is a region towards the back of the brain that has the important task of recognising the physical limits of the body. It gives us a picture of our personal boundaries – where I stop and the rest of the world starts – and helps us to orient ourselves with respect to our environment. Newberg discovered that in deep meditation, activity in that part of the brain is reduced. Coinciding with that reduction, meditators experience a loss of separation between themselves and their surroundings. At the same time that this is happening, blood flow is increasing significantly in another part of the cerebrum, the frontal lobes, which lie just behind the forehead and are concerned with focusing attention, in this case on the object of meditation. The double effect is an experience of the loss of a sense of self to be replaced – not by nothing – but by a greatly increased intensity of awareness of unity.

Newberg's objective was to see if it is possible to identify a natural process common to anyone following a broadly similar meditation routine, even within widely differing religious cultures. In his book on the biology of belief, he mentions that he was also able to persuade some Roman Catholic Franciscan nuns to take part in the same type of experiment, and though he does not go into detail it seems that he obtained broadly similar results. At the time of writing, Dr Mario Beauregard, a neurophysiologist at the University of Montreal, has recently reported on the results of a study that seems to complement Newberg's work.[28] Beauregard uses different techniques from Newberg, including functional Magnetic Resonance Imaging (fMRI) scans and Quantitative Electro-Encephalographs (QEEGs) of the brains of his subjects. They are 15 Carmelite nuns leading an enclosed and silent life dedicated to contemplative prayer, and he expresses his findings in terminology appropriate to that fact. His findings appear to complement and support the work of Newberg in coming to similar conclusions using different approaches:

- The experience of union with God (in the Christian sense) was multidimensional. The neural circuit correlated with this experience included brain regions normally involved in self-consciousness, the physiological and experiential aspects of emotion, sense of space, and mental imagery.
- These results suggest that the experience of union with God is not solely associated with the temporal lobe, i.e. there is no single 'God spot' in the brain.

WHY HAS OUR PHYSIOLOGY EVOLVED IN THIS DIRECTION?

Whatever final interpretation of spiritual experience one may come to, there is now enough research evidence to give conviction to the idea that it based on an inherited bodily process. This is entirely in tune with Alister Hardy's belief that it has evolved because it has survival value. We know from studying the archives in Hardy's unit (RERV) that people who report spiritual experience do find their life enhanced and that they are better able to cope with their difficulties. But how exactly does relational consciousness contribute to our survival? Can we make a general statement?

One clue comes from the work of the psychologist Michael C. Jackson who was for a time on the staff of the RERU in Oxford.[29] Jackson's unusual approach to this question was triggered by the ambivalence he had noticed amongst members of the psychiatric profession about the relationship between spiritual experience and mental illness. As we saw in Chapter 4, descriptions of intense spiritual awareness often resemble the diagnostic features of certain kinds of mental illness to be found in psychiatric handbooks. On this basis, some psychiatrists are inclined to interpret all claims to spiritual experience as indicative of pathology. Nevertheless, Jackson noticed a crucial difference between the two. Whilst sufferers from mental illness lose the ability to function properly in everyday life, people reporting religious experience almost always say that it was dramatically life-enhancing.

SCHIZOTYPY

Jackson decided to investigate two groups,[30] one consisting of people reporting religious or transcendent experiences, the other of people who had suffered from schizophrenia, but at the time of their interviews were in remission. He concluded that the similarities between the two forms of experience (which he labels 'p-s', representing 'psychiatric/spiritual') could be explained on the hypothesis that both are a function of what psychiatrists call a 'schizotypal' personality trait. He was already aware that diagnosed schizophrenics reported considerably higher rates of spiritual (and particularly 'numinous') experiences, than 'normals'. When he compared his two groups, he found that the onset of p-s experiences typically occurred during times of serious crisis in both the 'normals' and the 'schizophrenics'. Benign p-s experiences of the type reported by the 'normals' brought about effective solutions to these crises. The pathological experiences described by the 'schizophrenics' did not resolve problems in that direct way (though in the long term, some were felt to have had a beneficial effect).

Jackson's interpretation of the data is that transcendent experience is part of the normal experience of life, playing an adaptive role particularly at times of crisis. The process is triggered by high levels of personal stress, such as great fear, terror, loss of meaning and other extreme existential crises.[31] Mental illness results when this process is invoked but fails, increasing rather than reducing tension and producing a self-perpetuating circuit of psychotic episodes. This model differs fundamentally from most psychiatric models, in taking the view that such experience is positively valuable or meaningful for the individuals concerned. Jackson thus replaces a negative view of schizotypy with an explanation in tune with Hardy's hypothesis. On the basis of patients' reports he also accepts that there may even be positive spiritual aspects of full-blown schizophrenia. The same argument remains plausible when the data are considered from a Darwinian perspective. The persistence of schizophrenia has to be explained since, if (as appears to be the case) genetic factors are important in the transmission of the illness, why has it not been eliminated by natural selection? Jackson concludes that the general predisposition to p-s experiences is normally adaptive, and linked not only with problem solving but also with creativity.

Let's assume for the moment that schizotypy is indeed a helpful adaptation, as Jackson claims. How does it operate to solve existential problems? The work that Rebecca Nye and I completed on the spiritual life of young children tells us that it is likely to be something to do with relational consciousness. Formal religious practices like meditation or prayer permit us to become intensely aware that we are part of something that transcends ourselves. Most British people do not follow a programme of contemplative prayer, but we know from our large-scale surveys and from studying the archive of accounts in the Religious Experience Research Centre that this kind of altered awareness (the type investigated by Newberg) is triggered off spontaneously from time to time in most people's lives, especially in times of great stress.

This provisional account of the problem-solving process can therefore be put in terms of reduced alienation, the experience of coming back into contact with the fullness of relationship, after feeling abandoned and alone. It is important to note that people do not speak of a reduction of the all-too-real discomforts of existence – hunger, disease, violence against the person, ageing and dying. Spiritual awareness does not take those away. What it does do is make them more bearable because of the discovery that one's pain has a larger milieu. Particularly in monotheistic religions, the personal aspect is strongly stressed, and comes across as something like the transcendent God saying: 'I am with you completely; I am alongside you, sharing your suffering and helping to put it into a larger framework.' For members of religions without a personal God, the emphasis is similar; there is pain, but it is observed in the context of a monistic universe in which the self disappears in universal compassion, mercy and the other virtues.[32]

It is important to add that at the other end of the emotional scale, in an area of life not investigated by Jackson, moments of great joy seem to induce a similar activation.[33] This is quite clear from an examination of the archive, which contains many such examples. In either case, relational consciousness is experienced as drawing us into closer contact with the reality of our environment. It is the reverse of alienation. Spiritual insight shows us directly at the profoundest level that we are not isolated but deeply and inextricably continuous with manifold reality. Crucially, if we are to take the descriptions given to us seriously, we have to take account of the frequent remark that in the depths of spiritual awareness, the world is pervaded by a personal presence experienced as more vivid and intensely real than in everyday life.

THE BOUNDED AND THE UNBOUNDED SELF

Newberg's research on the changing roles of different parts of the brain during the course of meditation helps us to make physiological sense of what may be happening when there is arousal of relational consciousness. For most of our everyday TASKS the orientation provided by the activity of the parietal lobes is essential for successful practical interaction with our environment. At the same time the very fact of being aware of the boundaries of my body – that I stop at the surface of my skin and beyond that boundary is the 'not me' part of reality – reinforces my sense of individuality, and as will be made clear in the next chapter, is in tune with individualism, a philosophical doctrine that has had a major impact on European culture.

The key issue is this: giving priority to individualism is a matter of choice. The aspects of awareness I attend to, and those I ignore, are most likely to be guided by the culture into which I am inducted, though inborn temperament must also play a part. Within those parameters the criterion determining what I will focus upon is the aspect of reality that I judge to be important. Thus I may give great authority to the awareness of myself as isolated from the rest of reality and choose to ignore or suppress my sense of the continuum. But we have seen that separation is only one side of the coin of human experience. If I opt to shift my attention from the boundedness and isolation of my body towards an enhanced general awareness, as in the act of meditation or contemplative prayer, then what emerges is the perception that all things are bound together as One. In choosing to give importance to the meditative state, I focus on a different but just as real aspect of consciousness. In that form of awareness the difference between self and other is de-emphasised and my continuity with all of reality is made plain.

These findings of Newberg and others suggest that we can no longer take the traditional boundary of the self as given or obvious. Such an image is only part of the truth, overemphasised by the cultural history of which we are the inheritors. To ignore relational consciousness is to ignore a large part of what we are as human beings, with the negative social and political consequences I noted in earlier chapters. Holistic awareness is not often given high status in modern Western society, but this was by no means the case in the past. The relational consciousness associated with core religious practices was taken with the greatest seriousness. It was seen as offering the most fundamental insight into reality that is possible for the human species, as it still is in many other cultures.

RETURNING TO THE PROBLEM OF BIAS

It is an oft-repeated cliché that the scientific method deals with material reality and is not competent to make a judgement on the validity of philosophical or theological assertions. At the same time most researchers are very well aware that they bring to their work prior beliefs that affect the way they direct their investigations and interpret their data. Like most people trained in empirical science, I am a critical realist. I believe that our task as scientists is to strive towards an insight into the reality that presents itself to us, whilst remaining aware of our very real limitations. One of the (few) virtues of postmodernism is that it has made plain that complete scientific objectivity is an unattainable myth. The German philosopher Hans-Georg Gadamer pointed this out very clearly in his master-work *Truth and Method*[34] – no one is without prejudice. As there is no point from which we can view reality from the outside, we need to admit our partiality and use it constructively. What the reader of a scientific treatise *can* expect is integrity on the part of the researchers and openness about their presuppositions.

In the field that I have chosen to investigate, bias is a particularly challenging problem because the issues being explored are concerned with the profundities of human existence – matters of the greatest personal importance to people. In the previous chapter I discussed the contrasting views of Starbuck and Leuba on the psychology of religion. They were quite overt about their personal commitments, making it easy to examine the effect of their beliefs on the way they handled data. Today the situation is more complicated. Therefore I am concluding this chapter with a reflection on some of the ways that personal beliefs operate in people who currently work in this field, on the assumption that most of us are trying to be honest.

RELIGION AS EXPLANATORY OF NEUROPHYSIOLOGY

Someone who has a prior religious commitment is likely to be drawn to interpret scientific data in ways that add plausibility to their point of view. At what point is this legitimate, and where does it become unacceptable, at least scientifically? As my personal opinion is favourable towards religion, let me begin with what for me is a simple and obvious case of pro-religious bias affecting interpretation. Carol Albright and James Ashbrook's book, *Where God Lives in the Human Brain,*[35] claims to take an evolutionary stance. Each chapter looks at a particular part of the brain from a theological perspective. For example Chapter 5 is entitled, 'The Limbic System, Remembering, and a Meaning-Making God' and Chapter 7 is 'The Frontal Lobes, Intending and a Purposeful God'. One critic has remarked:

> I conclude that the image of God (imago dei) that the authors find reflected in the human mind/brain appears to derive from their Christian commitments rather than from evolutionary theory.[36]

My own interpretation of the book, when judged as a work of science, is the same. It might pass as a theological treatise. Perhaps when the authors describe it as a 'neurobiology of faith' they are stating just that. Even so, and though I share the authors' faith, I nevertheless felt uneasy when I was reading it because of a sense that the scientific data were being pushed illegitimately to make a theological point.

RELIGION AS A DISEASE

Religious believers are not the only people with boundary problems. The same issue arises for the sceptic who, having arrived at a personal conviction that religious belief is nonsensical, is inevitably drawn towards research that will help to explain how such a serious error can be so widespread. Intense dislike of religion both motivates creative research on the part of some biologists and leads others to step beyond strict scientific canons in their pronouncements. Towards the end of his life the Nobel prizewinner and co-discoverer of the structure of DNA, Sir Francis Crick, finally admitted publicly 'that his distaste for religion was one of his prime motives in the work that led to the discovery',[37] presumably on the assumption that demonstrating the chemical basis of living matter requires the repudiation of religious belief. That is an arguable point of view, but Crick went on to suggest that mystical experience might be caused by a 'theotoxin', literally a god-poison. Similarly, we have seen that Richard Dawkins (who *is* refreshingly direct about his personal opinion of religion) talks of religious memes as comparable to virus infections.[38] The evidence for either of these points of view or the appropriateness of such emotionally loaded language in scientific discourse is precisely zero, but the eminence of Crick and Dawkins has itself engendered a huge replication of the idea of religion as a contagious infection.[39]

RELIGION AS A USEFUL ERROR

Other less obviously biased hypotheses include a number of variations of the idea that spiritual awareness is structured into the biological makeup of the human species as a useful evolutionary accident. Apart from Persinger, the most prominent amongst those who have argued the case at length are the anthropologist Pascal Boyer[40] and the sociobiologist Scott Atran.[41] Atran and his colleague Ara Norenzayan summarise this kind of perspective as follows:

> Religion is not an evolutionary adaptation per se, but a recurring cultural by-product of the complex evolutionary landscape that sets cognitive, emotional, and material conditions for ordinary human interactions … Core religious beliefs minimally violate ordinary notions[42] about how the world is, with all of its inescapable problems, thus enabling people to imagine minimally impossible supernatural worlds that solve existential problems, including death and deception.[43]

Both Boyer and Atran in their different ways make out a detailed and plausible case for their arguments, including an acceptance that spiritual beliefs, though in their opinion nonsensical, do have personal importance to believers as well as evolutionary survival value. The credibility of their line of argument is based on a prior assumption that the entire religious/spiritual interpretation of reality is self-evidently mistaken. How can they be so sure that spiritual awareness has evolved by accident? Newberg's data, when combined with Jackson's suggestion about the problem-solving function of spiritual experience (i.e. through insight into the holistic perspective) offers a perfectly plausible hypothesis that requires no such leap of secular faith. Instead of dismissing folk narratives as nonsensical, a less loaded approach is to see them as socially constructed metaphors representing an underlying and essential reality of human experience.

RELIGION AS FALSE ATTRIBUTION

Attribution theory has a history going back at least to William James' *Varieties*, and it has been applied to spirituality by a number of students of religion. Proponents of attribution theory explain religious experience in the following way. A devout individual experiences an odd or unusual feeling, often in the context of a religious ritual or gathering. The feeling in itself has no special meaning, but it puzzles and perhaps alarms her. She casts around in her mind for an explanation, drawing upon her culturally available stock of meanings. In her case the cause of the experience is attributed to divine grace. That is to say there is a process of deduction leading to the attribution of a particular cause, but the deduction may bear no relation to the true state of affairs. The philosopher of religion Wayne Proudfoot's book *Religious Experience*[44] is probably the most impressively argued application of attribution theory to religion. Again there is an unstated assumption that a straightforward religious interpretation is a non-starter.

There are, however, reasons to doubt that attribution theory is adequate to the phenomenon. In particular, current developments in the study of emotion imply a rather more complex view of it than has hitherto been assumed. The role of traditional narrative and metaphor in meaning-making has been examined by Nina Azari, a psychologist at the University of Hawaii and her colleague Dieter

Birnbacher, Professor of Philosophy at Heinrich Heine University in Dusseldorf. Their subtle and philosophically informed discussion suggests that a major weakness of attribution theory is the avoidance of the possibility that we simply do possess intrinsic numinous awareness. Refusing to be open to that alternative is arguably a straightforward case of reductionism based on an inadequate understanding of the latest neurophysiological research on the relation between emotion and thinking.

Azari and Birnbacher make their case as follows: There are two traditional ways of understanding emotion:

1. Bodily or somatic theories that, in their extreme form, claim that emotion has nothing to do with thinking. On seeing a lion, I automatically have a bodily response immediately recognisable as fear, and there is no conscious interpretation involved. The best-known version of this view comes from William James, who applies it to religious experience. For him, religious feeling is immediate, known incorrigibly and entirely independent of thought.
2. Theories at the other extreme assume that emotion is reducible to thinking, so that even if there is bodily arousal involved, the specific emotion is the result of mental processing. I have a feeling and search through my repertoire of knowledge to determine what the feeling is. As we have just seen, Proudfoot is a leading representative of this perspective.

Neither of these approaches, say Azari and Birnbacher, stands up to scrutiny, especially in the light of recent neuroimaging research that demonstrates the complexity of neural processing during emotion. In the past, lack of knowledge of this complexity has led to a major philosophical error[45] in the interpretation of spiritual experience, i.e. the assertion that, 'Contrary to what the subject thinks, the object does not exist but is a delusion or illusion, wholly or partly caused by the emotion itself.'[46]

Both views fail to take account of what we now know about spiritual awareness from the studies of Newberg and others.[47] Spiritual experience certainly has a cognitive dimension that depends on learned (usually religious) beliefs acquired from the culture to which the person belongs. But these beliefs are not so much searched through like some kind of database to identify the 'cause' of the experience, as they are 'the field in which the experience operates'. The beliefs afford the space for the experience. Azari and Birnbacher sum up the point in the following way:

> The difference between the religious and non-religious view of the world is not a difference in factual information or factual expectations but in attitudes to the same class of facts. The dispute between the theist and the atheist is, unlike the dispute among scientists, not a dispute between rival hypotheses but between rival ways of seeing the world.[48]

To which I would add that compared to the (Western)[49] atheist's perspective, the religious way of seeing is much more immersed in the holistic insight derived from the physiological condition of the brain in deep meditation.

'THINKING THAT FEELS LIKE SOMETHING'

Recent empirical research on emotion is of considerable importance because of the ways that it offers new insight into theological and spiritual questions.[50] The British theologian John Bowker, like Azari and Birnbacher, reflects upon neurophysiological discoveries, in this case drawing upon the groundbreaking work of the Portuguese neurologist Antonio Damasio, currently head of the University of Southern California's Institute for the Neurological Study of Emotion and Creativity. Damasio has shown convincingly that there is an unbreakable link between thinking and feeling.[51] His finding undermines the assumption, current at least since the time of Descartes, that the mind is divorced from the body and bodily feeling. Thinking and feeling are inseparable, to the extent that standing back from feeling, in favour of cold detachment, is actually damaging to the full use of reason.

Feeling is not arbitrary; it has a directedness that comes from the fact that rational/emotional ways of responding to reality are genetically inherited from our parents. There is an identifiable anatomical source of these feeling responses in the region of the brain called the amygdala. In other words our biology sets limits to how we will relate to given aspects of environment. The significance of this is that it contradicts the relativist claim that we can make whatever world we like through social construction. There is a steadiness of perception – in this case of thinking/feeling – that binds our responses in to the given reality. The issue is quite clear from a Darwinian point of view. If we did not inherit skills that enable us to make very accurate assessments of the world as it is in its reality, natural selection would do its job and we would not survive.[52]

As an aspect of our physiology our spiritual experience too must have a similar, non-arbitrary thinking/feeling quality that enables us to make valid judgements about it, as indeed spiritual directors are attempting to do all the time in their role. This is not to dispute the importance of culture, which supplies the metaphors necessary for communication, but if the metaphors are adequate they are always created as responses within given biologically determined boundaries, shared by all members of the human species. That is why spiritually aware individuals are able to recognise common ground when they encounter the religious practices of another culture. John Bowker uses the term 'conducive property' to try to express this:

> ... it is now clear that we do make judgements, not just in science, but also in history, aesthetics and ethics, which are far from being subjective and relativistic in the extreme sense ... That is so because our brains and bodies are built in such a way that we perceive conducive properties in objects and people external to ourselves. These conducive properties evoke and sustain the judgements and vocabularies of satisfaction and dissatisfaction (*including the satisfaction of truth* [my italics]) and they do so with immense stability and consistency among humans even of different periods and cultures.[53]

In other words, our neurobiology ensures that subjective experience is by no means arbitrary, though it is expressed in thousands of different culturally determined guises. To use Azari and Birnbacher's expression, it is 'thinking that feels like something'.

The debates that I have been presenting have taken place in the context of a European history that, uniquely amongst human cultures, has constructed an extreme form of scepticism about spirituality. In the next chapter I will address the crucial question, 'Why is spirituality so difficult for Westerners, that is, people who have been educated within post-Enlightenment European thought?'

Chapter 9

WHY SPIRITUALITY IS DIFFICULT FOR WESTERNERS

Greed is good.

Gordon Gecko in the film *Wall Street* (1987)[1]
and
Billboard advertising Lynx deodorant (2005)[2]

We have to fight uphill to rediscover the obvious, to counteract the layers of suppression of the modern moral consciousness. It's a difficult thing to do.

Charles Taylor, *Sources of the Self*[3]

Not too long ago I spent several weeks taking part in a discussion forum on the Internet, run by the London-based *Independent* newspaper. One of the topics regularly debated was the plausibility of religious belief. Many of the participants were critical of religion, so I decided to submit the following question, 'Why do Europeans find religion so difficult?' For some people this was like waving a red rag to a bull. A small proportion of the replies were of the following type:

> As an adult with a job and a free thinking mind I have more important things to do than read through the babblings of a religiously impaired person such as yourself.

Or:

> I really can't see what all this crap or worshipping some ineffable being is about. Talking to people who believe in god, any of them, is like talking to children.

There were some people who also included at least a cursory reason for their annoyance. For example:

> It is a very straightforward matter to spot the primitive, infantile, pathetic mumbo-jumbo claims of religious ideologies of all kinds. How they offend so gravely against the well-attested findings of modern science and all good sense.

> Political thinkers from Machiavelli to Marx have seen religion as a means of control, and rightly so, since social control is actually the very essence of what any organised religion is.

> Difficult? Superfluous and damaging. Those are the words I would use for religion. There is not a religion in this world that does not write its history in the blood of innocent people.

And so on.

Now I know that the form of my question was provocative. Also, since those taking part in the forum had pseudonyms, there was no restraint upon people who enjoy unloading their anger anonymously. Nevertheless it seemed to me that the annoyance my question evoked was in itself a vivid symptom of the problem I was investigating. Religion is difficult for many people in our culture because (they say) it conflicts with the findings of science, because it is no more than a concealed means of social control and furthermore is a major source of violence and bloodshed.

These opinions are not new in Europe. In 1657 Blaise Pascal wrote as the opening aphorism of his notes for an apology for the Christian religion, later published as the *Pensées*:

> Men despise religion; they hate it, and fear it may be true.

Most historians of religion date the first emergence of widespread religious scepticism in Europe to the centuries surrounding the period when Pascal was writing, that is, the sixteenth and seventeenth centuries. But there is more to my curiosity. The complaints of my correspondents are familiar enough, and to a degree they are valid. Religion *has* been the cause of bloodshed; it *has* been used as a form of social control and it *has* been in conflict with science. These things are true of religion well beyond the bounds of Europe or Christianity. Yet the intense religious doubt that emerged in

European culture is distinctive. Europeans seem to have invented a very extreme form of religious scepticism – in a way that appears not to have been duplicated in any other culture.[4] That is a sweeping claim and my intention is to justify it by unpacking some of the factors lying behind the rise of individualism and, via its undermining of religion, its concomitant effect on spirituality.

THE SOCIAL DESTRUCTION OF SPIRITUALITY

Individualism has extremely complex roots in history[5] and since we Westerners are ourselves immersed in this history, its individualistic assumptions are likely to be hidden from us, unless we specifically look for them. When we do look closely it is possible to pick out five major steps in the construction of individualism. It is important to recognise that each of these steps has an ambivalent quality, that is to say, each is accompanied by both gains and losses for our humanity. Individualism is not the same thing as individuality, and the difference will become clear later in this chapter. For the moment it is sufficient to say that our individuality is what makes us unique, and to note the two earlier stages that are essential for this uniqueness to emerge.

1. HOW I COME TO THINK OF MYSELF AS 'I'

Words leave no fossils in the rocks, so the question of when language first appeared in the genus *Homo* is problematic. If we make a conservative estimate and confine language to our own species, *Homo sapiens*, then it goes back to perhaps 200,000 years at maximum. If on the other hand it is argued that the skills necessary for the manufacture of stone tools imply being able to speak, then this might apply to an older human species, *Homo habilis*, whose fossil remains in East Africa have been dated to two million years ago or more.[6] Either way, the effect of language on the construction of the self has been going on for a very long time.

Animals without language, although they are sensitive to their surroundings and relate to them in a knowing way, give at best only rather ambivalent indications of self-awareness.[7] Whilst they quite clearly have a memory, they lack the verbal apparatus for reflecting upon their memories or for considering the fact of their own existence. Consequently they live almost entirely in the here-and-now of the immediate events around them, immersed in an unbroken continuum that includes themselves. The distinction between self and other, though it is acted out in the way animals manipulate their environment, is never clearly articulated. We saw in an earlier chapter that the same is true of young infants. Adult observers of infants' behaviour can see quite easily that they operate in ways that implicitly recognise a distinction between self and other, as was remarkably shown in the work of Emese Nagy.[8] Putting it grammatically, they discriminate behaviourally between subject and object, but have little or no conscious awareness of the difference between the two.

With the coming of language a radical change occurs and by the age of eighteen months most healthy toddlers clearly and easily articulate the subject-object difference. The other job that language does is to provide a framework for memory and imagination. When we are able to name the things around us it makes them stand out for us in contrast to their surroundings and we can also reflect on them remotely, at other times and in other places.[9] One of the most prominent objects that an infant learns about through language is its self, a fact that is drawn attention to constantly by the parents when they teach the baby to say 'You', 'Me' and 'I'. The awareness of 'I' as an object of consciousness means that it can be thought about in the same way as any other object. 'I' begin to build up a set of memories and anticipations that make up a life history. I have become an individual. Nevertheless it is important not to lose sight of the fact that the process of individualisation is not done in isolation; it is always done in the context of a culture and in the company of others.

2. LITERACY

Compared to the span of existence of the human species, literacy arrived almost yesterday. Most people for most of human history have been illiterate and this was true until little more than a century ago even in the industrialised West. For example, in Britain it was not until the passing of W. E. Forster's 1870 Education Act that a concerted effort was finally made to eliminate illiteracy (a task that is not yet complete). Human consciousness has therefore evolved over many millennia in the absence of the ability to read and write. So in a way literacy is a move out of the natural and universal human condition. As we shall see in a moment, that natural condition still largely restricts awareness to the immediacy of the here-and-now.

It is hard for those of us who are able to read to imagine what it feels like to be illiterate, but a remarkable piece of pioneering research in the first half of the twentieth century allows us to have some insight. During the 1930s the Soviet government under the leadership of Stalin decreed the forced collectivisation of agriculture throughout the vast republic. The decree was combined with a vigorous effort to teach the peasantry to read and write because they needed to be literate if they were to be able to manage the complex work of the collective farms. A young psychologist at Moscow University, Alexander Luria, took the opportunity to study how the change affected hitherto illiterate peasants living in a group of remote mountain villages and pasturelands in Uzbekistan and Kyrgyzstan. These people were members of a primary oral culture, that is, they belonged to a community that had never been literate.

In summary, Luria[10] showed that, compared with people who could read and write, the thinking of illiterates is much more tied to the immediate situation (that is, the here-and-now) than to abstract reflections on the past and future. This meant that intellectual tasks which were elementary for literate people, for example simple classification, were difficult or impossible for them. In one of Luria's experiments, semi-educated and only recently literate collective farm activists were easily able to sort skeins of wool in terms of category: shades of blue, red, yellow and so on. On the other hand illiterate peasant women who, as expert embroiderers, were perfectly well aware of subtle variations of colour, usually named the skeins concretely, with terms like 'pig's dung', 'a lot of water', 'cotton in bloom', 'rotten teeth'. When asked to classify the colours into groups, for example shades of brown, the women would say things like, 'It can't be done. They're not at all alike; this is like calf's dung, this is like a peach.'

Luria tested illiterate men on their ability to complete simple syllogisms. One sequence went like this: In the North, all bears are white. Novaya Zemlya is in the North. What colour are the bears there? Most of the men were unable to give the correct answer, saying things like, 'How should I know, I've never been to the North. I've seen a black bear.' More crucially, people also seemed not to have much conception of themselves as individuals. For example, when asked questions such as 'What sort of person would you say you were?', illiterates were unable to describe themselves and suggested to

Luria and his assistants that they should ask someone else to answer for them.

Luria realised that such responses were not due to lack of intelligence, but to the structure imposed on thought by illiteracy. Literacy continues the process of individualisation initiated by the ability to speak, but with much greater impact. Literacy extends memory, permits us to classify and to generalise and gives us the ability to move in our imagination out of the concrete here-and-now and into lengthy abstraction. Above all, literacy opens the possibility of a private world and the ability to have a uniquely personal point of view, limited only by the size of one's library. In an important sense, literacy opens the door to personal freedom.

Complex modern society would be unable to operate without the skills that become possible with the ability to read and write. But the construction of a vast private world also potentially creates a blindness to our relationship with the here-and-now. Along with this loss there is the likelihood of a deterioration in our immediate sense of belonging to and being continuous with the surrounding community. In those traditional religious societies that are literate (Jews, Christians and Muslims are after all, 'People of the Book') this weakness is recognised and strategies have been created to counteract the loss of immediacy. Each of these cultures has developed a highly sophisticated set of practical exercises that help people to enter more and more deeply into holistic awareness of the here-and-now. I mean of course the skills of contemplative prayer (raising the heart and mind to God now, in this moment) or silent meditation (for example maintaining awareness of the act of breathing) currently being investigated by Newberg and others. These practices are undertaken by the faithful as a structured routine, often attended to several times each day. Ultimately the aim is to remain in this state of alertness permanently or, as St Paul put it, to pray without ceasing.

3. ABANDONING THE HOLISTIC PERSPECTIVE

What happens to consciousness though, when these strategies for staying in touch holistically are largely ignored, or abandoned altogether, as is more often than not the case in contemporary Western society? This constitutes the third step towards individualism. As literacy becomes more and more widespread it is more difficult, less natural, for people to enter the here-and-now awareness that is commonplace amongst members of primary oral cultures.[11] One effect on those who are highly literate is the increasing probability that they will acquire a disembodied, theoretical consciousness of the self, withdrawn from engagement in the surrounding environment.

The legacy in academic circles, perhaps especially in the field of empirical science in which I was educated, is an admiration for detached objectivity as a necessary professional stance. Like every other beginner in the laboratory, I learned that the inconstant and emotionally labile 'me' never puts water in a test tube. In writing up experiments 'it was noted' that 'water was placed in a test tube' by an abstract, clinically detached being who had nothing to do with the scruffy bunch of schoolboys occupying the classroom. This cult gave the false impression that human factors like hesitancy, error and free-floating imagination did not enter into the properly conducted research act. Taken far enough, training in detachment can include a distancing from other people and a loss of awareness of one's own emotional state. Intellectuals are notorious for 'living in their heads', that is, cut off from what is going on below the neck, sometimes to the detriment of their health.[12]

Those of us who have actually worked in a research lab know that Murphy's Law ('Anything that can go wrong, will go wrong') was probably discovered there, and that playfulness is the raw material of hypothesis construction. Nevertheless, objectivity in the sense of being able to stand back from one's work is important in scientific research and it is part of a much larger social movement. Many suggestions have been made about both the timing and the historical and political aspects of this growing sense of personal isolation. The nineteenth-century Swiss historian, Jacob Burckhardt, was one of the first to suggest a specific period in which individualism began to become dominant in European history. In his pioneering study *The Civilisation of the Renaissance in Italy*[13] he identified the emergence of the 'free person' as occurring first in Italy, pre-eminently in renaissance Florence. At the same time Burckhardt recognised that associated with this detachment was the appearance of a complementary private subjectivity, in that people recognised themselves as 'spiritual individuals'. We might guess that the process was not unlike the shift towards subjectivity witnessed by Alexander Luria amongst newly literate peasants in Uzbekistan.

Another suggested source of individualism is the Protestant branch of the Christian religion. The sociologist Max Weber claimed that Protestantism, especially in its Calvinist form, created an inner isolation in the believer sufficiently powerful to change the entire economic and political structure of the countries of the Reformation during the sixteenth century. This was brought about by Calvin's emphasis on the doctrine of predestination. Weber comments that anyone who takes this belief seriously is faced with an unprecedented inner loneliness:

> No one could help him. No priest, for the chosen one can understand the word of God only in his own heart. No sacraments, for though the sacraments had been ordained by God for the increase of His glory, and must hence be scrupulously observed, they are not a means to the attainment of grace, but only the subjective *externa subsidia* of faith. No Church, for though it was held that *extra ecclesiam nulla salus* in the sense that whoever kept away from the true Church could never belong to God's chosen band, nevertheless the membership of the external Church included the doomed ... Finally, even no God. For even Christ had died only for the elect ...[14]

The mere appearance of goodness is no guarantee, since anyone can make a public pretence of virtue whilst being inwardly corrupt. Therefore a robust doctrine of predestination encourages not only endless self-questioning, as Weber remarked, but also suspicion of the motives of others. A belief in predestination was not limited to Calvinism; it also appeared in certain seventeenth-century forms of Catholicism, especially Jansenism.[15] In his essay *Of Charity and Self Love*[16] written in 1674, Pierre Nicole, Jansenist priest and friend of Blaise Pascal explains how impersonation of virtue can be so accurate that it deceives everyone; hence it is not wise to trust anyone. It is deeply incongruous that a religious doctrine should have the effect of encouraging the erosion of the relational consciousness that underpins spirituality.

The idea of 'man alone' also gained currency in seventeenth-century Europe through the influence of the two dominant and contrasting philosophical perspectives of that period. The archetypal representatives are the idealist René Descartes and the materialist Thomas Hobbes. In the case of Descartes, his decision to make the

Cogito (Descartes' famous conclusion, '*I think*, therefore I am') the rock on which to build his philosophy had a devastating effect on the plausibility of relational consciousness. In the words of the Scottish twentieth-century philosopher John Macmurray,

> ... the adoption of the 'I think' as the centre of reference and starting-point of [...] philosophy makes it formally impossible to do justice to religious experience. For thought is inherently private; and any philosophy which takes its stand on the primacy of thought, which defines the Self as the Thinker, is committed formally to an extreme logical individualism. It is necessarily egocentric.[17]

At the other end of the scale, Thomas Hobbes' materialism was probably even more influential than Descartes' philosophy in promoting individualism. Hobbes was born in 1588 and lived through what historians see as one of the most violent periods of turmoil in European history. In particular the Thirty Years' War ravaged the continent throughout his early adult life. It is perhaps no surprise that he had a sceptical attitude towards the possibility of human benevolence. Most scholars believe he was a secret atheist at a time when publicly declared atheism would put a person in considerable personal danger.[18]

His materialist interpretation of human nature led him to the view that in the state of nature life is a warfare of all against all. If we cooperate with other people it is only because we see these interactions as in our interest (in this sense he was a precursor of modern biological theorists of reciprocal altruism and kin selection).[19] His assumption that each of us is in a struggle for power against everyone else is based on a materialist metaphysics stating that 'minds never meet, that ideas are never really shared and that each of us is always and finally isolated from every other individual'.[20] According to his most celebrated aphorism, life in the state of nature is 'solitary, nasty, brutish and short'.

People who have not read Hobbes are not always aware of the extreme violence he uses to describe the natural state of human society – totally at odds with the insights provided by relational consciousness. Thus:

> All men in the state of nature have a desire and will to hurt.[21]

In his master-work *Leviathan,* Hobbes makes explicit the brutality that people unleash upon each other in such a state:

> I put for a generall inclination of all mankind, a perpetuall and restlesse desire of Power after power, that ceaseth onely in Death … The way of one Competitor, to the attaining of his desire, is to kill, subdue, supplant, or repell the other.[22]

Here there is no law:

> To this warre of every man against every man, this also is consequent; that nothing can be Unjust. The notions of Right and Wrong, Justice and Injustice have there no place … It is consequent also to the same condition, that there be no Propriety, no Dominion, no Mine and Thine distinct; but onely that to be every mans that he can get; and for so long, as he can keep it.[23]

Hence the need for *Leviathan,* a Sovereign to subdue the anarchy and who himself gains that position through acts of terror or outright warfare:

> The attaining to this Soveraigne Power, is by two ways. One, by Natural force; as when a man maketh his children to submit themselves, and their children to his government, as being able to destroy them if they refuse; or by warre subdueth his enemies to his will, giving them their lives on that condition.[24]

After my reference in a previous chapter to modern research on the tender relationship between mother and child at birth, it is apposite to note that one critical contemporary of Hobbes said that he,

> … might as well tell us in plain termes, that all the obligation which a child hath to a parent, is because he did not take him by the heels and knock out his braines against the walls, so soon as he was born.[25]

4. INDIVIDUALISM UNDERPINS THE MARKET ECONOMY

The unbridled savagery that Hobbes loads onto human nature is of much more than antiquarian interest. The Canadian economic historian Brough Macpherson, one of the most eminent students of liberal democratic theory, asserted that Hobbes' account of society continues to dictate the organisation of the modern bureaucratic state. It is based, in Macpherson's phrase, on the doctrine of 'possessive individualism'.[26] The picture of human beings that Hobbes came up with was not simply the result of his free-ranging scholarly reflection. It was conditioned by the social order in which Hobbes was living; that is to say, seventeenth-century bourgeois society at the point where market forces first began to take on a dominant role. This is the fourth and most crucial step in the construction of European individualism.

No one supposes that selfishness was invented in the seventeenth century. Long before that, history was a catalogue of brutalities inflicted on others for the sake of avarice. What was new was the legitimation it gained at that point in time. Selfishness was to come to be seen as not merely acceptable, but a necessary expedient in the search for economic and political stability. The impassioned speech on behalf of greed as good by the reptilian financier Gordon Gekko in the film *Wall Street* takes its justification from this belief. In his essay *The Passions and the Interests*,[27] the economic historian Albert Hirschman meditates on the remarkable metamorphosis of the medieval sin of avarice into a necessary economic virtue. Dante's *Divine Comedy*, completed at the beginning of the fourteenth century, had envisioned sins of avarice as sufficient to consign their perpetrator to the fourth level of hell. By the end of the eighteenth century, avarice had come to be seen by economists as the pivot of the market.

Hirschman traces the evolution of meaning in the first place to:

> ... a feeling [that] arose in the Renaissance and became firm conviction during the seventeenth century that moralising philosophy and religious precept could no longer be trusted with restraining the destructive passions of men.[28]

Hobbes' initial solution, the advocacy of the straightforward repression of uncontrolled passion, came to be seen as inadequate. His pessimism about human motivation was not sufficiently responded to by the mere existence of a sovereign power. Who can predict if the sovereign will truly guard the peace of society, when in reality he may himself be a cruel despot, heedless of the cries of the oppressed, or merely weak?

According to Hirschman, the answer that emerged was to harness one of the passions against the others. The key to this solution, according to a whole series of seventeenth- and eighteenth-century thinkers, was the unquenchable desire for personal gain. The term that came to be used for this particular lust for possessions and which sanitised and set it apart from the others was 'interests':

> Because of the semantic drift of the term 'interests', the opposition between interests and passions could also mean and convey a different thought, much more startling in view of traditional values: namely, that one set of passions, hitherto variously known as greed, avarice, or love of lucre, could be usefully employed to oppose and bridle such other passions as ambition, lust for power, or sexual lust.[29]

The effect of this semantic drift is important, because throughout the seventeenth century, outside the field of economic and political writing, ordinary popular tracts on virtue continued to refer to avarice as one of the most repulsive of sins. On the other hand its synonym, 'interest', achieved a steadily enhanced status as the 'countervailing' passion. Finally, says Hirschman, it took on such a mantle of virtue, that in certain respects it was seen as more admirable, certainly more socially useful, than unselfishness. Thus in 1767 the Scottish economist Sir James Steuart could argue that in economic matters, self-interest is to be preferred to traditional virtue, *especially* a meddling concern for the public interest:

> ... were a people to become quite disinterested: there would be no possibility of governing them. Everyone might consider the interest of his country in a different light, and many might join in the ruin of it, by endeavoring to promote its advantages.[30]

The point was, as Steuart's colleague David Hume had also said of desire for gain, that it is a universal passion that operates at all times, in all places and upon everybody. It is thus much more predictable than other passions such as lust or revenge, which operate sporadically and are directed towards particular people. The very constancy of avarice had made it a virtue. Most famously, because of his influence on all subsequent economic thinking, in *The Wealth of Nations* published in 1776[31] the Scottish philosopher Adam Smith gave a financial rather than a political or moral justification for the unrestricted pursuit of personal gain.[32] Individualist philosophy (whether materialist or idealist) and the promotion of self-interest as the necessary basis for a stable market economy were mutually and powerfully reinforcing. They could not fail to be severely damaging to any trust in relational consciousness, and hence to spirituality. But beyond the four steps leading up to this point there is a fifth and final step to go.

5. RELATIONAL CONSCIOUSNESS TOTALLY REPUDIATED

We came across Ludwig Feuerbach in earlier chapters. When his book *The Essence of Christianity*[33] was first issued in 1841 it was greeted by many as a work of genius, but he was not to remain in that treasured position for long. He was soon to come under attack, first from Karl Marx, and then from Max Stirner, a rather odd and obscure member of the Young Hegelian group to which Marx belonged in Berlin.

Feuerbach is best known nowadays at second hand, via Marx's *Theses on Feuerbach.*[34] In this short series of 11 aphorisms, Marx takes him to task for failing to see that the religious illusion is created by unjust conditions in class society. In spite of Marx's criticisms, he accepted Feuerbach's suggestion that religion is a projection, and indeed Feuerbach was more thorough in his account of the process of projection than Marx, who simply assumed it to be the case.[35] Feuerbach's argument was built on a careful study of religious texts as well as noticing the way ordinary believers expressed themselves. He saw religion as a kind of secret anthropology. If you want to know about the best and highest qualities of the human species, you will find them by looking at the praises heaped by religious people upon God. Once we come to realise that the virtues attributed to God are simply projections of human virtues, we are able to emancipate ourselves from our religious delusions and replace them with ethical atheism. In other words a high-minded, noble morality is still the proper duty of humankind, but now cut away from its mistaken religious attribution.

One might feel that Feuerbach's repudiation of religion was as extreme as it is possible to get. Not so. His opinion was to be violently rejected as incomplete atheism by Stirner,[36] who preached outright egotism. With Stirner we see the final abandonment of any notion of relational consciousness. Stirner of course concurred with the rejection of a relationship with God as fantasy, but felt that Feuerbach was a sentimentalist who had failed to see the full implications of his discovery. Feuerbach, though a convinced atheist, continued to hold to the moral ideals advocated by Christianity. To Stirner such ideals were also projections, no different in kind from belief in God. For him all ideals and moral laws, without exception, are simply religion by another name, since they imply an imaginary and enslaving obligation beyond the self.

Published in 1845, four years after Feuerbach's *Essence of Christianity*, Stirner's only major work is *The Ego and Its Own*.[37] He currently appears to have something of a cult following, since his book has appeared in five different English editions over a period of 30 years, the most recent in 1995. At the time of writing, according to the Internet search engine Google there are more than 150,000 sites on the Internet relating to Stirner. Of all atheist writings, Stirner's is the most thoroughgoing in its uncompromising rejection of every philosophical, religious and political ideal, seen as nothing more than the depreciation of the individual:

> Away … with every concern that is not altogether my concern! You think that at least the 'good cause' must be my concern? What's good? What's bad? Why I myself am my concern, and I am neither good nor bad. Neither has meaning for me. The divine is God's concern: the human, man's. My concern is neither the divine nor the human, not the true, good, just, free etc, but is – unique, as I am unique. Nothing is more to me than myself![38]

And, reminiscent of Hobbes,

> For me you are nothing but my food, even as I am fed upon and turned to use by you. We have only one relation to each other, that of usableness, of utility, of use.[39]

Stirner's biographer R. W. K Paterson[40] comments,

> Whether owing to a failure of nerve, or to some basic astigmatism, the Feuerbachs and the Bauers[41] had all stopped short of the crucial point; at the last moment they had admitted the presence of some transcendental object in the scheme of things

> – not indeed a 'God' in the sense of a personal deity, but a 'Humanity', or a 'Society' or a 'Morality', all of which were as fictitious, and as autocratic in their claims upon the individual concrete human being, as any personal God had ever been; and thus the programme of atheism still remained to be carried through to its conclusion ...[42]

And with a brutality fully equal to Hobbes,

> Nothing, not even the primordial obligations not to lie, steal, kill etc. can induce the self-possessed egotist to take any step that is not in the fullest accord with his own distinct interests as he himself determines them ...[43]

Paterson sums up:

> Stirner's contribution to the German religious debate of the 1840s was to bring the whole debate to a momentary and stupefied halt. The full consequences of thoroughgoing atheism were now disclosed for all to see.[44]

Remarkably, Stirner's hero, the isolated self-sufficient individual, had already been identified and attacked ferociously by Karl Marx. He was none other than the capitalist fat cat, the unencumbered entrepreneur who is still with us today in plentiful supply. He is, in Marx's words:

> ... an individual separated from the community, withdrawn into himself, wholly preoccupied with his private interest and acting in accordance with his private caprice ... [for him] the only bond between men is natural necessity, need, and private interest.[45]

Marx was outraged by Stirner's book. In 1845 he and Friedrich Engels began writing what was meant to be a critical response to Feuerbach, entitled *The German Ideology*. But its main bulk turned into several hundred pages of attack on the follies of Stirner. It seems an obsessive response to someone whom Marx dismissed as small-fry. The part of the book devoted to Stirner is certainly turgid which is perhaps why it was not formally published until nearly ninety years later, in 1932.[46] Marx's disproportionate reaction has puzzled his biographers, who sometimes dismiss it as no more than youthful vitriolic exuberance.[47] It perhaps also suggests that, at some level, Marx realised that one line of materialist reasoning (leading back to Hobbes) could indeed eventuate in the disparagement of all ideas transcending the self, including Marx's own relational insight into 'man as a species-being'. This is a point I will return to in the final chapter.

Stirner's extreme individualism put into stark and uncompromising words what had been developing as an increasingly powerful, but muffled and disinfected assumption over the previous two centuries. Individualism encourages the complete suppression of relational consciousness and a consequent leeching away of ethical relations between the members of our modern commercial society. Once transcendence is abandoned (either belief in God or the kind of transcendental equivalent advocated by Feuerbach), morality becomes entirely subservient to what is financially prudent. In practice, Hobbes had already dispensed with all purposes apart from those that ensure the smooth working of the marketplace.[48] The binding obligation that remains in possessive market societies is to make sure the market does not collapse through financial mismanagement. In this circumstance the difference between moral obligation and what is financially prudent becomes insignificant.

Where financial prudence is the arbiter of conduct, politeness and care for the other person become suspect as no more than a manoeuvre, an optional extra to smooth the path of a financial transaction. In other words it is spiritually corrupt. Martin Buber makes the same point in his comments on Stirner:

> Responsibility presupposes one who addresses me primarily, that is, from a realm independent of myself, and to whom I am answerable. He addresses me about something that he has entrusted to me and that I am bound to take care of loyally. He addresses me from his trust and I respond in my loyalty or refuse to respond in my disloyalty, or I have fallen into disloyalty and wrestle free of it by the loyalty of the response … Where no primary address and claim can touch me, for everything is 'My property', responsibility has become a phantom …[49]

The difficulty for Stirner is that he has entirely lost touch with relational consciousness. For Buber, he is a sociopath:

> He simply does not know what of elemental reality lies between life and life, he does not know the mysteries of address and answer, claim and disclaim, word and response ...[50]

Macpherson notes a yet more ominous difficulty at the heart of individualism. Paradoxically, individualism *needs* collectivism and, as he puts it, 'the more thoroughgoing the individualism, the more absolute the collectivism', as in the totalitarian authority of Hobbes' *Leviathan.* Totalitarianism is the logical endpoint for a society that has lost touch with relational consciousness. Buber makes the same point in his insight into the role of extreme individualism in the creation of the authoritarianisms of the political Right and Left.

> 'True is what is mine' are formulas which forecast a congealing of the soul unsuspected by Stirner in all his rhetorical assurance. But also many a rigid collective We set in, which rejects a superior authority, is easily understood as a translation from the speech of the Unique One into that of the it which acknowledges nothing but itself – carried out against Stirner's intention, who hotly opposes any plural version.[51]

Nazism and Stalinism are prominent examples of the genre, but the category should also include some forms of Western religious fundamentalism, certainly those that have come under the influence of Calvinism. In their polarisation of the world into good and evil empires and their disregard for human life, they have become tragically disengaged from the relational consciousness that underlies genuine religion. The political consequences of this dismissal of spiritual awareness are very great, and around us in full measure at the present time.

CONCLUSION

Stirner and Hobbes between them bracket a period in European history marked by the progressive and cumulative discrediting of a fundamental aspect of our biological make-up, relational consciousness. The casualties of the process are all around us. One example that became salient during the second half of the twentieth century is a decline in social participation throughout the Western world. This has been comprehensively documented for the United States in Robert Putnam's study, *Bowling Alone.*[52] He provides statistics that show a collapse since the 1960s across almost all social behaviour: political and civic participation, informal social groupings, altruism, volunteering and philanthropy, reciprocity, honesty and trust. The loss of what Putnam calls 'social capital' is graphically illustrated in the ubiquitous deployment of the paraphernalia of surveillance (cameras, electronic tracking devices, alarm systems, databases) to discourage crime. They are a totalitarian means of controlling a society in which relational consciousness no longer forms the basis of a moral commonwealth.[53]

I have indicated my belief that the source of these discomforts is located in long-term social processes first detectable in seventeenth-century Europe. Their appearance coincides with the rise of religious scepticism.[54] From the perspective I have presented, this loss of formal belief has behind it, at a more fundamental level, a suppression or repression of relational consciousness, because of the individualistic requirements of the marketplace. Accordingly, the Jesuit historian Michael Buckley[55] identifies a change in direction of religious apologetics at this period, away from an appeal to personal experience of relationship with transcendence and towards the argument from design. As we saw in Chapter 2 the design argument as deployed by William Paley was still dominant 200 years later, when Darwin was a student in Cambridge.

Theologically, the move amounted to a long-term abandonment of the sense of personal relationship with the immanent God, and replaced it with an intellectual conviction based on detached philosophical argument. Given the individualist assumptions of both materialist and idealist philosophers at the time, this was an understandable strategy. But what was being discarded was any sense of trust in the relational consciousness that I have argued is an inbuilt feature of the human organism.

It is no wonder that spirituality has an increasingly difficult time in the Western world, and along with it the plausibility of religion. Commercial and intellectual pressures force us towards a heartless individualism that cancels relational consciousness out of the human equation. Nevertheless, the empirical evidence I have discussed strongly suggests that spiritual awareness is a constant in our biological makeup. The fact that human decency and mutual trust continues to be widespread is evidence of its resilience, even though severely constricted in its range by the straitjacket of individualism.

Furthermore, since individualism is quite clearly a socially constructed ideology, there is always the possibility of deconstruction. In the final part of this book I will return first of all to our Nottingham non-churchgoers to hear more of their diagnoses of our spiritual distress, now seen more directly at the institutional level. In the last chapter, responding to their advice and the empirical evidence I have presented, I consider some possible ways to make our spiritual tradition young again.

Part 4

FACING THE CRISIS

Chapter 10

THE PROBLEMS OF THE INSTITUTION

I sign the book, donate an Irish sixpence.
Reflect that the place was not worth stopping for.
Yet stop I did: in fact I often do,
And always end much at a loss like this,
Wondering what to look for.

Philip Larkin[1]

This book has been concerned with spirituality, not religion, but one thing has become crystal clear during our investigation. Spirituality is about relationship. It is about being profoundly in communion, body and soul, with the totality into which we find ourselves 'thrown'. When that communion becomes attenuated or broken and we are left to our own devices, we become reduced and personally impoverished, concerned first and foremost with how to manipulate the world to accommodate our desires. From Thomas Hobbes to Max Stirner and beyond, there have always been advocates of the conviction that all I have is myself. Hobbes at least saw the murderous consequences of his belief and began the process that still continues, of setting up defences to protect the physical integrity of the commercial arena in which we are forced to compete.

Hobbes' version of individualism leads to the politics of despair. When taken to its limit, it destroys true community and replaces it with the environmental destruction and totalitarian nightmares that have marred history. There is nothing new about individualism as such, but it is hugely powerful in our time because it is built in, almost as an axiom, to the economic system that governs the global market. Fortunately, rapidly accumulating scientific evidence suggests that spiritual awareness is a permanent feature of our biology, potentially subverting and countermanding the dictates of avarice. The huge upsurge of concern with spirituality currently taking place throughout the Western world implies its indestructibility and is a sign of hope.

The problem is that very often it is an isolated, secret or privatised spirituality. Disconnected relational consciousness is a contradiction in terms and its isolation means that it is permanently in danger of being suppressed or even repressed in the face of a dominant secularism. At best it survives as the private consolation of the individual, one of the paradoxical consequences of the decline of genuine community. Until such time as we recover a public spirituality, vast numbers of people are condemned to loneliness of spirit. Furthermore, privatised spirituality has no political purchase and therefore cannot be brought to bear with full power on the structural evils of the day.

The social representation of the spiritual life in the Western world has overwhelmingly been via the religious institutions, and I as a Christian believer remain convinced that the churches hold the key to spiritual recovery. So what kind of response to the spiritual revolution might we hope for from the churches, to which many references have been made in earlier chapters? I want to return to the opinions of our Nottingham interviewees, now listening specifically to their views about the institutions and their official representatives. Their advice will be important when, in the final chapter, I come to consider ways of bringing the churches to life.

THREE WAYS OF RELATING TO THE INSTITUTION

As you would expect in a group of people specifically selected because they do not go to church, the commonest feelings to emerge were negative. Sometimes they were extremely hostile. But a deeper immersion in the data shows that under the cloak of anger there were a number of tensions and polar oppositions. These tensions suggest that much of the fury directed towards the churches arises from disappointment.

All who took part in the research conversations had either abandoned membership of the church or had never even considered it. The images of God and the interpretations of reality purveyed by the religious institution simply did not resonate with their life experiences. Hence the churches appeared to have little or no bearing on the content of their spiritual lives.

Even so, in some cases the church was tolerated as a necessary inconvenience. In spite of people's reservations, it represented dimensions of life still considered important: morality, a sense of the past, groundedness, even of 'belonging' or identity. This does not mean that the individuals either needed or wanted to be a regular part of the church community themselves. It was enough to have the children christened ('given the right start') and perhaps attend a service at Christmas. Anything more tended to be viewed as excessive. In summary, the ways of relating to the institution fell approximately into three categories (illustrated by the case studies in Chapter 3):

- *Believing/not belonging.* This group of people tended to be in the over-40 age range. Usually they had quite a good knowledge of Christian belief and doctrine, but for various reasons had ceased going to church. Their spirituality was conventionally Christian.
- *Not believing/not belonging.* Here, people had been brought up with quite a lot of contact with the Christian institution. Their experience of it had been negative and they were generally quite hostile towards it. Nevertheless they had an easily recognisable spirituality, sometimes of considerable depth. On the whole the members of this group came from the younger age range.
- *Untouched by the church.* This group of people appeared to have had no significant contact with the religious institution, yet they often had a vivid spirituality. In some ways they were the most interesting people we met since they had perforce to construct a personal spiritual doctrine, drawn from a variety of sources.

The binary opposition that was present in much of what people said was between the notion of belonging (the church is 'my tribe'), and the church as alien, not allowing me to belong. The 'tribal' end of this tension was held to most strongly amongst people who had had a Roman Catholic upbringing, but, rather to our surprise, was also true, to a degree, even of people who had never had any links with the institution, that is, the members of the third category above. However, some people's experience when they did venture into their local parish church meant that they felt strongly that they did not belong, as in the instance mentioned below of Sharon's embarrassing visit to a communion service.

CYNICISM ABOUT THE RELIGIOUS INSTITUTIONS

Cynicism about the religious institutions was the 'default mode' in public discourse, as was made very clear in the focus groups when people were asked for their opinion of the church. In fact the references to hypocrisy, bigotry, being out of touch and other critical clichés were boring in their repetitiousness. Whether these criticisms are justified or not is beside the point. They express vividly and frankly a wide range of perceptions of the religious institutions by a group of people who have chosen to distance themselves from them. Critiques of the institutions fell into the following categories:

RELIGIOUS INSTITUTIONS FOSTER IGNORANCE

Not many members of the groups held to the classical secularist view that religious belief arises from ignorance, but Paul was one such person. He was a true agnostic with regard to belief in God in the sense that he hummed and hawed about it. Nevertheless he inclined to the view that religion stems from an unawareness of natural causes, aided and abetted by the (fortunately declining) power of the institution:

> I think religion has come from our ancestors not having an understanding of how the world evolves, and how everything comes about. And that is, if one year you don't have a rainfall and your crops don't grow and your friends are dying and there's famine, then they've turned to a rain god ... and it's extended from there somehow. And I think religion will go on, because as long as we live on this planet I don't think anybody will be able to say how the earth was created ... People will carry on being religious because they just do not know.

RELIGIOUS INSTITUTIONS ARE RIGID AND AUTHORITARIAN

There was a strong sense that the religious institution dissuades people from thinking for themselves. Joanne contrasted religion unfavourably with spirituality. Whereas spiritual people 'think a lot about lots of things', religious people 'just base their thoughts on traditional religious type things'. Sometimes it seemed that people were browbeaten into conformity. For a brief period Belinda had a Catholic boyfriend and used to go to church with him:

> They just do go to church every Sunday and that's it. When I had my brief association it was rammed down your throat rather. I can remember sitting in the church and the priest shouting from the pulpit – it has stayed with me you know – banging his fist on the thing and 'you will do this' and 'you won't do that' ... and nobody dared breathe without his permission. It was quite daunting really.

Belinda wanted to add that she had never experienced that kind of behaviour in the Church of England and had found vicars to be 'most approachable, very nice people'. Nevertheless, according to Carol, Anglican clergy have their rigidities, teaching doctrine of which they can have no direct knowledge. She had been reflecting on how nobody knows what happens to us after we die. So,

> How can vicars stand up and say that's going to happen, how does he know any more than anybody else? ... It's all the things he's learnt over the years ... he's read it and he believes it you know.

Sarah had encountered what she felt was intimidation when she had tried to have her daughter christened:

> But the priest from the church came round to see me while I was pregnant and he sort of put me off really, he said that because I got married in a registry office, I wasn't actually married in the eyes of God. And he also said, 'Don't think you can just come along to the church and get your child christened, you know, it just doesn't work like that.' I felt he was very intimidating ... so I thought well, don't think I'll bother.

James contrasted the loving presence of which he was deeply aware in his meditation with what he had been taught in his religious education:

> The sort of doctrines and things I'd grown up and been educated into were about some authoritarian figure with rules and regulations ... there's a distinction between personal direct

> experience of this Other ... maybe we could say God, and the kind of teachings that one has as a child. The problem about those is [that] I had ethical and personal objections to what was being said, and anyway it doesn't correspond to my experience of this Other.

James had come to see adherence to a set of rigidly imposed rules as an evasion of personal responsibility. For him, formally religious people were often approval-seeking, wanting 'to be seen to be doing the right thing'.

James was hinting that religion encourages what psychologists call an 'external locus of control'. That is to say, instead of finding authority for their personal behaviour and beliefs from their internal resources (having an 'internal locus of control'), some people – and according to James this is true of people drawn to religion – prefer to follow rules set for them by an outside authority. Some such idea may lie behind Emma's remark that the churches 'pick on vulnerable people'.

RELIGIOUS INSTITUTIONS ARE NARROW-MINDED

Emma felt that belonging to a church limits you by keeping you out of the real world. Matthew and Paul independently spoke of their awareness of the overwhelming mysteriousness of existence, a mystery that was trivialised by the unreflecting 'standard answers' often purveyed by the religious institutions. These simplistic certainties have another unpleasant consequence, according to Matthew. He felt they had the effect of tearing apart the seamless relationship within the human community:

> That's a problem I do have with anyone that has certainty; that faith [to them] means surety and certainty. And that I think prevents objectivity. Because you believe this set of things you therefore cannot entertain something else that is happening simultaneously can also be true.

He saw 'certainty' as creating categories of people who by definition are outsiders:

> An easy sort of thing to point out would be an inner-city Islamic School or a Catholic School where they're ringfencing this set of beliefs and it therefore means that that grows stronger and the people outside of that become more inhuman,

> and more remote, and not integrated ... By reinforcing the beliefs within that group you are distancing yourself all ways, you're going in the other direction, aren't you?

Paradoxically, Matthew recognised the need for certainty in himself, but stressed how it must be allied to openness:

> We all of us need to cling on to certain private certainties and, well, certainties that are shared by a group that we wish to be associated with. But you can find toleration ... and entertain the possibility that what you believe can at any point be challenged, but is nevertheless important, and viable and truthful to yourself.

But some of the certainties clung to by religious fundamentalists were just too foolish for Matthew to countenance, quite apart from functioning to exclude people. Such beliefs were not merely silly, they were dangerous because they led to exclusiveness and the rejection of people who do not fit, the despised and rejected sectors of human society.

In line with his insight into the complexity of the world, Matthew added,

> I realise we have to put a shape and face to him, don't we, I suppose? It's not on to just be happy with something being amorphous.

The pigeonholing of Jesus in this way nevertheless seemed blasphemous to Mary:

> There wasn't anybody he ignored, the lame, the dirty people, the clean people, the rich people, the poor people, the prostitutes, lepers. He embraced everybody, but in Christianity we tend to think of Christ as being white, and clean and probably middle-class.

Religious belief was commonly equated with being socially conservative. Graham saw this as paradoxical. He could not understand how religious believers could in good conscience support the death penalty, except as an expression of thoughtless social inertia. He also saw the row over women's ordination in the Church of England in that light:

> That was for me so hypocritical, that they were more or less saying man is better than woman; [that] we can do a better job than what they can. I thought that dreadful.

RELIGIOUS INSTITUTIONS ARE HYPOCRITICAL

This was the commonest allegation made against both churches and churchgoers. Sometimes this meant a critique of double standards on the part of churchgoers, for example the contrast between a cloying 'niceness' (Matthew revealingly called it 'humble arrogance') in the context of the church, allied to disgraceful behaviour in personal and professional life outside the church. Double standards sometimes extended inside the church doors. One woman spoke of feeling unable to go to church because her husband had been mentally ill and she feared the gossip of her fellow churchgoers.

James finally turned away from his devout childhood when he came across double standards amongst outwardly devout students when he was studying theology at university:

> I felt among the Scripture Union types there was a lot of superficiality and dishonesty and I didn't want that to be a part of my life. [They] were always making sure they were at all the prayer meetings and going to church three times on a Sunday, who really really couldn't give a damn about their fellow man or woman.

Paul laughingly remarked, 'Be a car salesman five days a week, ripping people off and then go to church and make it up on a Sunday'. Alan put it this way:

> There seems to be a lot of hypocrisy in the church. You see such an awful lot of old people who put their Sunday best on and went out ... [and] didn't always practise what they preached on the other six days of the week. It seemed to be a lot of older people repenting before they passed on.

More subtle, but in its way just as damning, was a difficult-to-pin-down feeling of insincerity emanating from some church people, which came across as 'syrupy piety'. At one time Emma had had a great deal to do with the church, but looking back she thought 'a lot of it is very false, very forced'. Robert gave this as a major reason why he didn't like being in church:

> I thought there was a lot of false people there. There was a lot of people being nice to people, acting in a certain way, because they thought they should be. And you know it really winds me up.

This lack of integrity extended beyond parochial life into the media. Joan had worked for a television company on religious programmes, but remarked that she 'wasn't that impressed by the people that were doing it as totally sincere'. In general the portrayal of religion on the media did it no favours. Sean dismissed TV religion as boring. James felt religious broadcasts usually chimed in with his personal experience of religious people:

> They come across as symbols of hypocrisy, [though] I think at Christmas time carols can be quite nice to listen to … from Cambridge. I might watch it for a couple of minutes before I change the channel.

Simon came across as a fairly ordinary working-class man. Although he did not attend church services himself, he had a deeply felt belief that life unfolds in a meaningful way. He had a heartfelt longing for faith and respected the genuineness of his wife's religious practice, to the extent of giving a lot of practical help in the church. Yet as far as he was concerned, the loving community showed him precious little love:

> What puts me off when I go down to the church … there's people that I have known from there over ten years and they don't speak to you. Right? Then I can walk up the lane and they go down to the church and I can walk by them and they don't speak to you … If you don't talk with a plum in your mouth, if you don't [have] the right clothing you're not entertained, are you? That's not polite, that's not good manners, nothing. I mean being a Christian, you're supposed to be open, friendly, like the person my wife is.

Simon's anger was made absolutely clear towards the end of the conversation:

> I mean like if my children grow up like some of the people I've sat with, I think I'll have done a bad job.

RELIGIOUS INSTITUTIONS DAMAGE PEOPLE

Paul's mother and brother were deeply involved in the life of a quasi-Christian sect, but he had distanced himself from their beliefs. He offered the observation that 'there are a lot of nasty religious people about'. He later expanded on this point by asserting that, 'the biggest murderer of people has been religion, religious wars, or in the name of religion'. Perhaps reflecting the Protestant history of England, Roman Catholicism was picked out by several people as particularly unpleasant. In Carol's opinion the Catholic members of her focus group 'had had a lot of religion from young' and seemed to have been 'brainwashed'. Similarly, Matthew felt strongly about Catholics in his group who had expressed uneasiness about lapsing from their faith:

> ... the damage the Catholic Church does with burdening its beliefs with guilt about all sorts of things. It's fascist you know, it really is ... they're doing that to people and messing their heads up like that, and they weren't in their early twenties you know.

Simon was deeply distressed by the way the Catholic Church treated his mother:

> ... my mum was a Catholic. She committed suicide, which I could relate to because she [had] an illness. Like when I was going through the process of getting her buried the Catholic Church didn't want to bury her. 'Course that really did me, 'cos my mum did a hell of a lot for the church, like my wife does now. And you think, how can people treat people like that?

John's estrangement from the Church could be traced back to his childhood and,

> ... this sort of argument my mother had with the formal Church and the Catholic Church saying to her 'If you don't bring up your children to be Catholics, and you don't find good Catholic godparents for them, they'll spend eternity in Purgatory,' and all the rest of it. I suppose, you know, that to me seems like brainwashing ... So I developed my own connection with God.

Rather more mildly, Steven, whose wife is a practising Catholic, felt that the parish priest was keeping an eye on the regularity of her church attendance,

> People weren't as often at church as they can or the priest would like and I think they did a few home visits. We didn't get one mind you, which I thought perhaps we should have done, but perhaps my wife was worried about getting one, you know.

Emma was brought up as a devout Baptist and retained many conventional religious opinions, including a literal belief in hell as a place of eternal fire. But she was scandalised by an encounter with a brutal theology when she was eleven years old. She had been very distressed when her cousin had died at the age of six and remembered asking the minister, 'She will go to heaven won't she?' He said, 'Well, did she go to church? Was she a God-fearing child?' Emma said, 'Well no she didn't, her parents didn't'. 'Well no, then', he said. Emma added,

> I suppose then, at eleven being told that your six-year-old cousin is going through that – that to me is very, very wrong and shouldn't happen. And I don't believe it does happen.

Emma pondered the tension between an institution she saw as corrupt and its role in mediating the faith, following a comment that in spite of her distress she still seemed to have quite strong religious beliefs,

> Well I think being brought up around it as well … oh, no, that's contradicting myself slightly, but yes, I do have strong beliefs. I don't know quite where they have come from as such …

Lucy retained some remnants of religious faith in spite of her experience of attending a Pentecostal church in her childhood:

> I hated being in the church. All the people in there made me feel really uncomfortable … and there was like speaking in tongues as well … I really just didn't understand what was happening … it was almost like it twisted around what you had already learned … It took all the happiness out of it and made it a scary thing.

Jenny's experience of physical brutality from her preacher father was a major contribution to her desire to leave the church:

> There was a lot of hypocrisy I really couldn't cope with because my father was quite violent towards me. I did find that incredibly difficult to cope with, the fact that he was going to church on a Sunday, and standing up in a pulpit, and preaching to people, and then coming home and being absolutely awful to his family.

She now felt some compassion for her father, seeing his religiosity as a means of coping with weakness:

> I began to understand him a bit more over the years. A lot of it was out of his control. And maybe that was one of the reasons he felt the need to go to church himself, because he couldn't cope with what he was being.

Where criticism of the church carried particular weight was when it came from someone who, though they had ceased to attend church, continued to be personally devout. Mary, whose devotional life was thoroughly Christian (remember how she spoke of her daily prayer, and how through prayer she had learned of the compassion of God: 'God is not a snooper') enumerated the defects of the church as:

- obsession with control, (the church should be a servant);
- living in the past (God is a God of the living, not the dead);
- failure to be concerned with humanity as a whole – meaning universalism, and genuine political commitment to the poor (occasional collections for the poor of the world are mere tokenism!).

Unlike many others, Mary was not hostile to religion, but saw these defects as a betrayal of religion, enough to make her avoid church attendance altogether. She angrily pointed out that stopping going to church has nothing to do with losing faith, 'we need more religion now, not less'.

ATTITUDES TO THE BIBLE

The orthodox faith of the 'believing/not belonging' group I mentioned above implies knowledge of the Bible. Nevertheless, their response to questions about the Bible, like everyone else's, was remarkably thin. Not even one of the participants spontaneously mentioned that they currently read it. It was definitely not an open, or opened, book. On the whole, vague memories of Sunday School, catechism, or RE classes were all that people could contribute to their knowledge of the contents of the Bible. It seemed to have no clear connection with most people's personal beliefs or understanding of God, though undoubtedly the language they used could often be traced back to the Bible. But this seemed to be primarily because of the cultural inertia to which I referred in Chapter 4; biblical language and references have diffused widely into common speech over the centuries. Where people did care to respond directly to questions about it, their comments hardly strayed beyond the sentimental view that the Bible contained some 'nice stories'. Emma with her devout Baptist upbringing was one of the few who expressed a more sophisticated view, having studied St Luke's Gospel up to examination level at school. Nevertheless, when asked whether it figured in her current life she said,

> No, I don't read the Bible. I mean my experience of the Bible would be purely from my churchy type days ... I sometimes wonder why people read it, because I couldn't make head nor tail of it. And I also believe it is quite outdated. It was written by a man, a lot of it, and therefore it's not the be all and end all ... I mean it's like saying, do I read Shakespeare, and you know I don't read that either.

Probably the most positive comment came from Matthew. His deep interest in spiritual matters meant that he took religious culture seriously but added, with reference to the inerrancy of Scripture, 'We shouldn't have to believe in these things verbatim.' Yet,

> ... we've had two thousand years of this, and apart from anything else, it's an immense part of our social history, and it's important to all of us, believers and non-believers and if we are non-believers we should be allowed to have a stake in it.

FEELINGS ABOUT CHURCH BUILDINGS

Our conversations showed that there are plenty of sacred spaces available for people, even in the secular world. They are those corners of life where contemplation can take place. They can be as varied as sitting by a back doorstep in the evening, relaxing in a swimming pool on a Sunday morning, playing a quiet round of golf, enjoying the open countryside, or even visiting a garden centre. But traditional sacred spaces such as churches or cathedrals still have a powerful attraction for many who choose to stand outside the institution. There is a striking contrast between people's attitudes to religious doctrine and religious people, which as we have seen carry a heavy burden of criticism, and their positive feelings about religious buildings. (We have already seen this in the case of Tom, discussed in Chapter 3.)

Church buildings were still typically seen as belonging to everybody and open to all. That is to say, they are not merely the property of churchgoers. Individuals spoke of going into empty churches and appreciating the ambience. Somehow there was a different quality to the atmosphere in a church compared with other, secular buildings, however grand. You will recall that Matthew said his interest was artistic, but his remark that churches have 'weight and silence' seems to carry something more than mere aestheticism.

Graham wanted to be sure we understood that he was not very religious, and yet,

> Now as I've said, I'm not particularly religious. I don't go to church. But the feeling of calmness inside there [Ripon Cathedral] and the feeling of humbleness if you like, you know, is, it was amazing … It was a strange, strange feeling – strange feeling. To say that I don't really feel particularly religious, it was, it was a calming, calming atmosphere.

Churches sometimes provide a safe haven when life is particularly difficult. The following quotation comes from Simon, whose two-year-old nephew died in tragic circumstances. Feeling very upset, he found himself making his way into his local parish church:

> They've got a stained glass window straight in front of the altar. And I don't know what it was, but when I walked out of there I felt a hundred per cent better. Just being there. I don't know if … I don't know, I can't explain it. I haven't got the words to explain it, but I can definitely say I felt a hundred per cent as I walked out after that, about the events.

Simon had difficulty in articulating his experience. This echoes

Graham's 'strange feeling'. Being in a church is somehow different to being in other buildings, but the difference is almost inexpressible. Belinda said she couldn't put it into words, that somehow she felt she was entering a parallel universe, then added:

> My son had to do a project on his local church and I went with him ... When you walk through the door there's a feeling of tranquillity and it's just peaceful ... it's almost like everything's slowed down. It's a bit like when I go away on holiday and I sit on the beach ... there's something about the sea, [it] brings everything closer somehow and slows everything down.

Evelyn visited the medieval Minster in Southwell with one of her children:

> I am overawed by places like that and I do feel a kind of peace when I walk in them ... It's a house of God and no harm can come to me there. It's just a place of sanctuary. I don't feel that way if I walk round a stately home, you know. I think, 'Well this is a lovely building' but it doesn't stir anything in me like that. I don't think 'I'm OK here.'

However, churches are not always viewed so positively. Sarah felt rejected by the Catholic Church because of certain life choices she had made, and the actual church building seems to symbolise her feelings towards the institution:

> It's huge, it's enormous, it's like, I don't know why they built it so big, but you go in and it's like the ceilings are massive, it's freezing cold, and like it's so big, they don't fill it ... I think it's just too big, it just feels like cold in there in the winter, you're all wrapped up, and can't wait to get home. That's not how it should be, is it?

CONNECTING PERSONALLY WITH REPRESENTATIVES OF THE INSTITUTION

There was a great deal of reticence about talking to the clergy about religious matters. The shyness was related to an almost universally expressed fear. To open a conversation with a minister of religion would invite an embarrassing attempt to guide them into accepting Christian beliefs about which they were at best doubtful. People in general were put off by their perception of religious orthodoxy as requiring them to sign up to a list of beliefs. This came to a head on those occasions where contact with the church was sought, for example because of a wish to have an infant baptised. People's experiences were strongly contrasting, either extremely positive or extremely negative and awkward, depending on the reception they obtained from the clergy. They became very angry when they felt that they were being kept out of the church. They also sometimes felt they were being asked to jump through hoops just to keep the clergy happy. However, if parishes were welcoming, then people spoke warmly of their experiences.

Failure or refusal to make a connection with the religious institution was in part related to the gap in people's heads between 'religion' and 'spirituality', to the detriment of the former. For Carol religion is about rules:

> 'Religious' means that you go to church very regularly and do all the things. Like certain churches have different rules and you abide by all the rules – like Catholics go to confession and this, that and the other.

From this perspective Carol saw the churches as rather like well-regulated social clubs for the elderly, giving them the chance to get out of the house, since 'probably they've not really spoken to anybody all week'. In contrast to this instrumental function, she felt that 'spirituality' has real profundity, because it lies at the root of ethics. It is:

> … an awareness of knowing that there's something and you don't really know what … there's kind of an awareness that you yourself … would never hurt anybody, or do anything to anybody else.

In spite of (or perhaps because of) her utilitarian view of the institution, Carol was married in church and the service was conducted by 'a lovely man … who didn't force us to go to church or anything else … He knew we obviously believed something and he was happy with that.' It was another story when she approached the church to have her child christened, because the vicar required her to give a public undertaking to attend the weekly service, something she had no intention of doing:

> It was horrendous. I just wanted to curl up and die. It was really quite nasty. He kept implying that unless you went to church there was no point in getting them christened … and I'm thinking God, I've got to face these people.

Tracy and her husband Ron wanted their children to go to a church school. Ron said:

> You have your kids christened, you know. I mean, we had to have the kids christened to go to the school we want them to go to, because the main thing is the kids and education. I mean, it's a good school.

Perhaps embarrassed by Ron's frankness, Tracy wanted to be clear that instrumentalism wasn't the whole story, adding,

> I don't think that was the reason why I wanted to have them christened particularly. Obviously now you look back and you think, well it's probably helped with the schools and things.

And it is true that a number of people spoke of their desire for the ritual of baptism for their offspring, even though they felt remote from or awkward in relation to the doctrinal background. Carol thought christening was important for her children: 'I just wanted them to have that and then they could do whatever they like when they get older.' There are questions to be asked here about the function of ritual as something rather more profound than the mere enactment of a superstition.

Alan had a strong feeling for 'rites of passage' and thought that christening was 'getting off on the right foot'. He was amazed that some people do not go through these rituals:

> To me it's like a natural thing. It's like I said, you get married, you have children, you go to church for the christenings; you die, you go to church for the funeral. It's just part of that process.

Apart from christenings, weddings and funerals, Christmas was the one occasion in the year when a significant number of the people to whom we spoke admitted to having attended church at least on one or two occasions. This was in spite of their assertion that they were not churchgoers. Christingle services were particularly popular with families, and many people mentioned enjoying singing hymns and carols. This seems in the main to be homesickness for a (perhaps fictitious) past when all was well with the world. People spoke of Christmas having 'lost its meaning' for society, but within the walls of the local parish church there remains some memory of the way things used to be. Here again two perspectives are in tension with each other. For these people the church is attractive because it represents the very unchanging certainties that, at other points in their conversation, are criticised for their rigidity.

James had a cynical view of church attendance as 'a kind of display for an hour on Sunday'. But beyond the shallowness of display, or Utopian longing, or loneliness, there was often a perceptible desire to belong to a spiritually aware community such as the church 'ought' to be providing. In the case of Matthew (and some others) this was not currently an option because he believed he would be required to sign up to a set of beliefs about which he was, at best, dubious.

A few people, especially those who had had a strong religious upbringing, saw themselves returning to the fold at some future date. Sarah, with a devout Catholic background, equivocated, 'I want to, I might, I will go.' Joan, though expressing her distaste for street evangelists and 'happy-clappy' religion, expressed her certainty that there is 'something there'. She was disappointed that she had never had a visit from a clergyman because, 'I would really like to get back into going to church.'

Many of the younger members of the focus groups were very remote from the institution. If they did become curious this sometimes led to toe-curling embarrassment. Sharon remembered one occasion when she ventured into the local parish church and felt terribly uncomfortable:

> But I think they ought to do like a church for beginners really, because if you're not used to going, because they always have communion here. [She goes on to explain how she was encouraged to go forward for communion.] It was a really awkward situation; do you know what I mean? And he was giving us the sip of the wine, and the um, and he beckoned us to bring the children up as well, and they give you, whatever it is they give you to eat. Is it rice paper?

As I have just noted there often seemed to be no place for the 'beginner' and equally as serious, no place for the honest searcher. One of the most moving occasions during our research conversations was when people, usually towards the end of the chat, spoke of their personal search, and many of them were on the search. Colin was burned up with anger at the institutional church, yet later on spoke wistfully of his longing for a plausible basis for belief. Another man explained sadly that he had been unable to find a church open enough to accept him as a searcher.

In the face of what I have presented – a barrage of criticism mixed with nostalgia, especially for the physical presence of church buildings – are there ways of treating the institutional sickness diagnosed by our interviewees? This is the subject of the final chapter.

Chapter 11

TREATING THE SICKNESS OF THE SPIRIT

As with Schopenhauer, Beckett came to the conclusion that life is no blessing and it would have been better never to have been born.

John Calder[1]

Nothing human is alien to me.

Karl Marx[2]

Church people drive too slow.

Quote from the film *The Big Fish*[3]

DIAGNOSIS OF A SPIRITLESS SITUATION

I began with Samuel Beckett and I want to finish with him. Commenting on the 2005 revival of *Waiting for Godot* by its original director Peter Hall, the theatre critic Michael Billington said:

> Fifty years after *Godot*'s turbulent London premiere, a Bath audience sits through it with rapt attentiveness and gives it a rousing reception – not, you feel, because they have been told it's a classic, but because they intuitively recognise Beckett's portrait of a world without hope or resolution.[4]

Beckett's personal response, according to his biographers,[5] was stoical decency, laced with humour and a refusal to let pessimism get in the way of his stubborn sense of moral responsibility. And who would want to argue with that? But although he diagnosed our 'sickness unto death',[6] he offered no cure. As far as Beckett was concerned, he was describing brute reality for which there is no remedy, though we repeatedly find ourselves in a state of dependency, hoping that someone will save us. To put the point frankly, Beckett has nothing more constructive to say to us because he himself is trapped in a closed circuit of hopelessness.

Karl Marx's meticulously researched and impressive investigation of the same spiritless situation did offer a programme of treatment, though few now think the practical outcome has been anything other than disastrous. The problem is related to a somewhat obscure ambiguity in the way Marx used the term 'species-being'. It is important to make the point clear, because it affects how Marx's followers have understood his model of human nature. I noted that in the 1844 manuscripts he revised Feuerbach's picture of species-being as a kind of inbuilt human solidarity (or relational consciousness) by adding a dynamic, socially constructed aspect.[7] Nancy Bancroft focused on this when she commented that, 'We differ from the beasts in that our individual and *species character is not given at birth*' [my italics].[8] This interpretation contains a familiar and significant uncertainty as to whether, in highlighting social construction, Marx implicitly abandoned the belief that species-being is also an aspect of our primordial human essence, i.e. that there is an inbuilt ethical dimension to human nature.

Certainly he more or less completely dropped the term 'species-being', following the publication of Stirner's *The Ego and Its Own* in 1845.[9] The point is this: if species-being is not inbuilt, and if you accept Feuerbach's argument about the origin of religion (and Stirner certainly did accept it), then you are surely bound to recognise that high-minded rhetoric about one's moral duty is in the same bracket as pious religious talk. When you put yourself under an obligation to be self-sacrificing, or feel a duty to others, you are not responding to anything real; you are in servitude to a myth with origins no different than those of religion. Nicholas Lobkowicz summarises the dilemma created for Marx:

> If he could not predict the proletarian revolution in terms of historical necessities entirely independent of philosophical speculations and ideals, Marx had to defend his own ideals against Stirner's attack. And this he was neither willing to do nor capable of doing.[10]

The effect of the change in Marx's language was to blot out his original notion of ethical awareness as built in and replace it with historical inevitability. The criterion on which to judge the correctness of an action became more and more whether it fitted the impersonal onward roll of the dialectic of history. In the Marxist regimes that began to appear in the twentieth century this was to lead to widespread self-sacrifices on the altar of what was itself a new Feuerbach-style myth, 'historicism'.[11] To a great degree the blame can be put on the double-talk that Marx indulged in following Stirner's demolition of Feuerbach. The demagogues who contributed to the totalitarian horrors of many so-called Marxist regimes should have paid more attention to the humanity of their mentor, whose favourite maxim was *nihil humani alienum puto* (nothing human is alien to me).[12]

The failures of the Marxist governments in the Warsaw Pact countries are events of huge and direct political significance for the market economies of the West.

Whilst the popular explanation for the collapse is in terms of material failure (they couldn't compete with capitalism), the deeper cause was an ethical and spiritual blindness, curiously reminiscent of the collapse of virtue into financial prudence in the Hobbesian marketplace. For the market economies of the West ignore the spiritual dimension of what it is to be human, just as much as any Marxist regime. Possessive individualism and Stirner-influenced amoral Marxism have the same origin: the sixteenth-century materialism best represented by Thomas Hobbes.

Marx was very properly outraged by the collusion of the religious institution with the ruling class of his time. Yet hidden within that institution was an essential truth not at all uncongenial to him, which emerged during the 1980s, a spirituality of human solidarity, or as the Poles called it, *Solidarnosc.* Vaclav Havel, who was instrumental in bringing about the regime change in the former Czechoslovakia, put the argument succinctly:

> Reasonable commerce between human beings presupposes behaviour that is in accord with what human beings actually are. When this is not so, rebellion eventually occurs at some deep level.

Havel's critique is equally a warning to the individualistic West.

The sheer success of the market economies in maintaining and increasing material living standards has until recently obscured their defective model of what it is to be human. But now, perhaps for the reason given by Havel, there is a new recognition of spirituality sweeping right across the Western world. Unfortunately it is an ambivalent phenomenon. As I noted in the previous chapter, spirituality that buys in to the individualism of the surrounding secular world condemns itself to being self-contradictory, superstitious and vulnerable to fanaticism. The absence of a shared community can even mean that spiritual experience is not recognised for what it is. It becomes a private possession, or an ego trip, because individualist assumptions lead to a failure to recognise its universal implications. Such self-centred spirituality, if it acknowledges a relational dimension, may limit itself to a sense of belonging to the sub-group in which it occurred, for example one of a multitude of tribalisms – narrow political, ethnic and sporting groups, or fundamentalist religion. The universalism inherent in spiritual insight collapses under the pressure to set up a boundary (cf. the Berlin Wall, Belfast Peace Line, Israeli security fence, apartheid, segregated football stands) between the in-crowd and those who are outside. At a still more extreme degree, the experience of unity may become perverted into huge self-aggrandisement and in the end take on a demonic quality, as David Tacey has remarked.[13]

Using the language of Emmanuel Levinas,[14] the privatising of spiritual awareness makes it easier to lose touch with the 'face' of the other and hence of the sense of unconditional obligation. What happens to relational consciousness is illustrated by the comments of our Nottingham interviewees. Almost without exception they expressed a persistent feeling that the environment is not as trustworthy as it once was; that mutual obligation is disappearing. People repeatedly talked of shrinking back into the safety of the home and in the process contributing to an erosion of relational consciousness or what some writers are beginning to call 'spiritual capital'.[15] Until such time as we recover a plausible and publicly available spirituality, vast numbers of people are condemned to a loss of spiritual capital or even to spiritual destitution.

In summary, when relational consciousness loses its public voice, it not only restricts its range, it also fatally weakens its ethical[16] and political impact. The community accordingly suffers great damage, not so much from overt crime as, in Marx's language, alienation. Its members are adrift in Samuel Beckett's wasteland.

RECOVERING THE PROPHETIC ROLE OF THE RELIGIOUS INSTITUTIONS

That, I assume, is where institutional religion comes in, or should do. But the accumulated research findings I have reported overwhelmingly confirm the negative image of the institutions. We have seen that our interviewees told us in no uncertain terms that the religious institutions we have inherited in the West[17] contain an immense baggage that no longer corresponds to people's personal experience. Hence, in the terms of David Tacey, they have been 'mercilessly discarded, especially by the young'.[18] How then, can we go about renewing the institutions so as to reconnect them with the burgeoning but individualistic spiritual upsurge of our time?

The main part of this closing chapter is a reflection based (a) on the advice implicitly given to us by our interviewees, (b) on the insights that modern empirical research has opened up for an understanding of the embodied aspect of spiritual awareness, and (c) the impact of a dream I had during a period of intense concentration on what our research findings implied. In my dream I was seated in church when suddenly the whole of the West wall collapsed, leaving the building wide open to the outside world. Within moments the division between inside and out disappeared and people of every kind were moving easily in and out. I awoke with a feeling of relief, for it seemed that an unnecessary boundary had been swept aside.

1. LEARNING TO ENTER THE HOLISTIC MIND-SET

In devout communities contemplative prayer forms the major framework of daily life, with times set aside for it throughout the day and sometimes during the night. Sufficient empirical research has now been done to make it clear that important physiological changes take place in people who enter that mode. Thus we have seen that the research of Andrew Newberg and others demonstrates that meditation is associated with observable and measurable alterations in the neurophysiology of the brain. So far, the cross-cultural evidence suggests that whatever the cultural context of contemplative activity, it opens up awareness of participation in, or relationship to, the totality. That is to say, it positively engenders relational consciousness.

It is embarrassing to repeat what must sound like a platitude, but the recovery of contemplative prayer really is the primary task of the currently enfeebled institutions. My personal observation as a churchgoer is that in many churches genuine prayer is given lip service but is not always practised in reality. A background of understanding of the contemplative mode is implied during congregational participation in the ordinary spoken prayers during a religious service. Whether those prayers are extemporary or taken from a set prayer book, it is assumed that the members of the congregation have placed themselves in the presence of God. That is to say, by an act of recollection they have entered a mode of here-and-now awareness with which they are already familiar. But if they do not already engage in regular contemplative prayer they are quite likely to have limited insight into what is meant in practice and to make the prayer little more than a rigmarole of religious words.

Petitionary prayers that are read from a text may have little or nothing to do with a genuine raising of the heart and mind to God. Sometimes this is because they are hurried or read without awareness, or because the words themselves are a distraction. At evening prayer in my schooldays the headmaster frequently repeated a petition that began 'O Master Carpenter, who through wood and nails hast fashioned our salvation'. He obviously liked the juxtaposition, as I did, but enjoying a clever figure of speech is not the same thing as prayer. Similarly, extemporary prayer is not always what it seems. The old joke about the Edinburgh minister who begins his prayer with, 'O Lord, as thou hast probably read in *The Scotsman* the other day', mocks the habit of addressing prayer to the congregation rather than to God. As the Jesuit theologian Karl Rahner pointed out, God-talk is not necessarily equated with genuine spirituality, nor does spirituality always involve using the word 'God'.[19]

These commonplace aberrations are a world away from true contemplation and one might feel that encouraging holistic awareness is an uphill struggle. There are at least two reasons for optimism. Both involve helping people to recognise that they already know the contemplative mode, and then making a connection with the liturgy of the church:

- Firstly we have seen from our research that spontaneous moments of spiritual awareness are a feature of the lives of most people from time to time, so the area of human experience is at least familiar in that sense. This is especially the case when someone is going through a personal crisis (severe illness, loss of a loved one, loss of livelihood). On these occasions secularist assumptions do a poor job in helping to cope with the trauma. People quietly drop their culturally learned reservations, permitting the opening out of an awareness of transcendence. As we have noted, this opening out can also be triggered at times of great joy. Drawing attention to these moments and associating them with formal prayer is a valuable initial orienting exercise.

- Secondly, the realm of spiritual awareness is by no means remote from contemporary life, where it features in the ever-expanding field of humanistic psychology. All of those approaches to personal growth that emphasise bodily awareness have the purpose of assisting a recovery of the 'point mode', which as we saw is also the mode of contemplative prayer. Mainstream counselling methods like those of Carl Rogers,[20] developed for use either in a one-to-one situation or in a group, depend upon the ability to bring one's awareness into the here-and-now. This is also the case with the widely used Tavistock model for studying the group dynamics of organisations.[21] The therapeutic work of Eugene Gendlin,[22] which depends upon the focusing of awareness on immediate bodily feelings, is another example, and is particularly interesting because his method has been adapted for use as a religious exercise by two Catholic priests, Peter Campbell and Ed McMahon.[23] The work of the sociologist Alfred Schutz[24] on what he calls 'tuning' and Mihaly Csikszentmihalyi's studies of 'flow'[25] both have direct relevance to the prayer mode. The American sociologists Mary Jo Nietz and James Spickard expressly refer to Schutz and Csikszentmihalyi in their efforts to develop a sociology of religious experience.[26] In addition, the Spiritual Exercises of Ignatius Loyola have been picked out by Isabella Csikszentmihalyi[27] as an example of an attempt to create a state of 'flow' for the whole of a lifetime. Ignatius' exercises also have contemporary secular counterparts in the Italian psychotherapist Roberto Assagioli's *Psychosynthesis*[28] as well as in the gestalt approach to psychotherapy developed by Frederick Perls and others.[29]

The existence of these and other similar practices in the field of humanistic psychology reminds us that even in secular life the contemplative mode is by no means an archaeological leftover. The training of clergy and others charged with spiritual education ought to include careful investigation of these connections, so central to all religious insight and so frequently neglected.

2. BECOMING CONSCIOUS OF HOW METAPHOR SHAPES EXPERIENCE

One way of thinking of cultures is to see them as rather like valves that either open up or close our contact with aspects of reality. The crudest way they do this is by the use of force. Throughout history and most likely in every society, certain kinds of knowledge have been forbidden as either hostile to the dominant belief system, or blasphemous, improper, or a danger to peace.[30] In medieval Europe the desire for knowledge itself was looked upon with a certain amount of suspicion, based on St Paul's remark, 'Knowledge puffeth up but charity edifieth',[31] and too much curious investigation of nature was felt to have 'something of the serpent'. Religious censorship at a certain period inhibited the growth of empirical science and certainly damaged the lives of individual scientists, most famously, Galileo. Unfortunately there are still cases of legal sanctions closing off knowledge in the name of religion. One well-known example is the attempt of some states in the Bible belt of the USA to curb the teaching of Darwinian ideas in schools.

Equally prominent in our own day is the use of legislation in an attempt to restrict our awareness of the spiritual dimension of our experience. The open censorship and physical persecution of religious believers in the former Soviet Union is well documented[32] and similar repression is currently operating in North Korea and China. The problem for the authorities is that legal sanctions are a crude and only partly effective way of suppressing thought, because they arouse resentment and clandestine resistance. Resistance can be crushed, but history shows that countercultures can also survive and even flourish, as was the case with Roman Catholicism in Poland under Communist rule.

There is however a subtler, more powerful way in which the social valve closes down alternative ways of thinking. It is through metaphor. Metaphors are much more than rhetorical devices to make language more interesting. They shape the way we think about reality and how we act, and because they suffuse the whole of language, for most of the time we are unaware that this is happening. Lakoff and Johnson in their excellent book on metaphor[33] give the example of the way we often portray 'time' as money:

> You're wasting my time.
> This gadget will *save* you hours.
> How do you *spend* your time these days?
> I've *invested* a lot of time in her.
> You need to *budget* your time.
> He's living on borrowed time.[34]

And so on. To use this metaphor is not just a colourful way to talk about time; it has the power to shape the way we handle our time, perhaps by making us anxious and fearful of being 'spendthrift'. A different set of metaphors can change behaviour. For example, describing time as 'a vast ocean' or 'an infinite expanse' is likely to be accompanied by a more relaxed mood.

Since spiritual experience is often claimed to be ineffable, we would expect metaphor to play a large part in religious language, and in recent years academic theologians have written a great deal about this theme.[35] Our research demonstrates unequivocally that the metaphor systems commonly used in this area are problematic for the ordinary person in the street. In contrast, parallel studies of the use of metaphor in empirical science show that it is highly regarded as having a key role in fostering creativity and hypothesis construction.[36] Why then are religious figures of speech often felt to be so remote that they appear to belong to the realm of fairy tales? The difficulties people speak about are the dissonant relationship of religious metaphors with current social norms and their remoteness from contemporary cosmology.

One example of dissonance comes from a biblical metaphor system highly appropriate to the economy of ancient Israel, based on the importance of sheep herding. People concerned with mission must have very quickly encountered difficulties with these images in regions where people do not keep sheep. The anthropologist Edward Evans-Pritchard writes of one comical dilemma:

> I have read somewhere of the predicament of missionaries to

> the Eskimos in trying to render into their tongue the word 'lamb', as in the sentence 'Feed my lambs'. You can, of course, render it by reference to some animal with which the Eskimos are acquainted, by saying, for instance, 'Feed my seals', but clearly if you do so you replace the representation of what a lamb was for a Hebrew shepherd by that of what a seal may be to an Eskimo.[37]

In the New Testament, Jesus describes himself as the 'Good Shepherd', who in turn cares for the human family, seen also as 'sheep'. In today's world the connotation of people as 'sheep' is highly objectionable as an epithet because it implies weakness and blind following, as secular critics have been quick to point out.[38]

The most clichéd of all religious figures of speech, the idea of God as an old man in the sky, illustrates the use of an outdated cosmology. On the occasions when our interviewees had got as far as attending a church service they certainly encountered it during the recitation of the Creed, with its reference to Christ's Ascension into heaven to sit at the right hand of God the Father. Almost all those we spoke with were perfectly clear that anthropomorphic imagery is not to be taken literally, but the fact that the metaphor is drawn from a world picture discarded by Copernicus nearly five hundred years ago means that it is strongly associated with the past, not the present. The splendour of God enthroned in heaven, towering over the universe in the manner of William Blake's *Ancient of Days* is magnificent, but turns very easily into the cartoon image of a bad-tempered pensioner mouthing imprecations whilst seated on a cloud.

The lampooning rejection of this metaphor is also due to the experience of a number of our informants that traditional images of divine transcendence are intolerable to them. In particular, the men in the sample had assimilated a picture of God the Father as a terrifying, punitive 'superego' figure, remote from their own experience (the alleged Freudian origin of this image has been the subject of a number of psychological investigations).[39] On the other hand many of our informants had been personally aware of God's immanence, a loving presence 'closer to us than our breath' or 'as near as the jugular vein', and these metaphors were accordingly much more acceptable.[40] At times the disjunction between immanence and transcendence leads to the two being torn apart and apparently seen as separate gods, as in the case of our Nottingham informant James discussed in a previous chapter. James's ideas came out of a profound personal reflection that led him to seek the God beyond the conventional God.

In other cases the way people talked of God suggested the alienation Feuerbach talks about. Their picture of immanence came across as theoretical rather than based on practical experience; a sentimental and cosy projection of God that is reassuring when life becomes pressured. This kind of metaphor is linked with an understanding of the function of the church as above all to offer comfort. The insinuation is that the primary task of spirituality is to minister to people who are personally inadequate and to soothe the elderly as they face death, hence keying in to a major contemporary reductionist explanation of religion. In a previous chapter I mentioned the 'deprivation' theory[41] which proposes that people are religious because they are lacking in some way (i.e. because of poverty, loneliness, ill-health, social repression, etc.). Normal people, the implication runs, don't need that kind of crutch. That in itself is destructive of any kind of image of the church as challenging or heroic and it must be said that the notion that religion could be demanding did not appear very often in our conversations.

This suggests to us that the religious institutions, in so far as they have reached these people, have failed to offer a profound enough understanding of the spiritual search, or the function of metaphor. Our non-churchgoers are implicitly inviting the churches to join with them in exploring spiritual depth but find they are palmed off with superficiality. In view of all that has been said of metaphor above, a major investigation is necessary to discover in detail the nature of the cultural disjunctions that dismay a sometimes suspicious or hostile public. This does not necessarily imply abandoning traditional imagery, although that might seem advisable in the case of some figures of speech that have become extremely remote from contemporary reality.

3. WEAKENING THE POWER OF METAPHORS THAT DAMAGE SPIRITUALITY

Let me briefly return now to a metaphor that *is* taken seriously and is positively destructive of spiritual awareness. Following on from what I said in a previous chapter, it is clear that our economic system is built on the assumption that an all-pervading aspect of human nature is possessive individualism. We saw that most economists since the eighteenth century have operated on the axiom that human beings are motivated above all by self-interest. This is particularly true of those countries that dominate the global market. From their restricted point of view the picture of human beings as self-interested consumers of commodities is an optimistic metaphor because self-interest, it has been asserted since the days of Adam Smith, is necessary for the market to work efficiently. The picture of us as possessive individualists destroys the plausibility of more communally minded metaphors – for example seeing ourselves as 'species-beings', or our fellow human beings as 'our neighbours' or our 'brothers and sisters in God'. Such talk is in danger of being an embarrassment or dismissed as sentimental nonsense. ('Why does the idea of service to others sound so wimpish?' a friend of mine asked the other day.)

What I have been saying points to the fact that the European cultural valve has closed off the full story of our biological makeup. Can anything be done to retrieve our lost humanity? I believe there are numerous possibilities and they depend on two assertions:

- The empirically supported claim that spiritual awareness is a biological reality and not just a socially constructed fantasy. When it becomes known, the scientific evidence now accumulating on spirituality is itself emancipatory and removes some of the taboo.
- That there is a feedback effect between developing spiritual awareness and the motivation to build a society rich in social capital. They mutually encourage each other and this leads us to consider ways of catalysing communal awareness.

The religious institutions are potentially in a very strong position with regard to social experiments in community-building. More than any other large body, the church has vast experience of the practicalities of community-building, through two thousand years of sponsoring the monastic orders. Medieval ideas of community are not ours, but contemporary adaptations suggest feasible ways forward. The following example of a modern experiment in community building is particularly interesting because it takes the spiritual dimension with great seriousness. It is drawn from the work of the late Douglas Trotter.[42]

Trotter was for many years an academic theologian, but prior to that he was on the Staff of the Pioneer Health Centre in Peckham in London. The 'Peckham Experiment'[43] was an attempt to realise the ideal of a positive approach to health in the broadest sense, or as Trotter calls it, 'being fully alive'. Founded by George Scott Williamson in a small way in 1926, the Centre took a holistic stance, characterised by the metaphor that, as the body is the ultimate significance of the cell, so perhaps we might consider the human being as having the same positive and benevolent relationship with the cosmos. In line with this view the aim of the Centre was not merely to deal with illness but to encourage abundant life through giving great emphasis to the building of community. The Centre at its height consisted of a club of some thousand families living within walking distance of the architect-designed building and it can be seen as a forerunner of many subsequent experiments in holistic medicine.

Trotter's idea grew out of his early experiences at Peckham. In his book *Wholeness and Holiness*[44] he reflects as a theologian on society as it is, and as it could be. Having worked in the field of holistic medicine he uses his experience to draw a parallel between our spiritual life and our physical health. The desiccated state of contemporary medical science, he says, has analogies with the deadening of our spiritual life. He sees the British National Health Service, which should be concerned with abundant life, as no more than a 'sickness service' primarily focused on the avoidance of death. Spirituality in Western Christianity is in the same state. Consequently the discipline of self-denial – originally understood as a joyful response to the experience of God's nurturing Spirit – has been displaced by a grim-faced obedience to laws, duties, and regulations; indeed, much as Stirner envisaged them. The metaphors applied to the spiritual life are filled with negativity:

> Thus, the 'way of salvation' is no longer viewed positively as the realisation of human potential through the liberation of [people's] natural powers but negatively, as a way of combating the perversions of [our] created nature. It is sins-oriented and may be described as a 'fight against sins'.[45]

The aridity Trotter rails against is inevitable in a society where people accept the individualist metaphor and its associated family of sub-metaphors ('Look after Number One', 'Business is business', 'Get the other fellow before he gets you', etc). In that case the general assumption must be that human relationships are at bottom a technical, mechanistic matter and the notion of true intersubjectivity is no more than a myth we cling to for comfort.

In response to this shared medical and spiritual crisis, Trotter proposes a method for retrieving insight through a theory of 'correspondence' between the realm of direct intuitive knowing, where spiritual awareness lies, and the physical world, where knowledge is acquired primarily via the methods of empirical science. Trotter pins the blame for the crisis in the Western church on the disconnection of theology from its natural and living source in human experience. He turns for inspiration to the Orthodox Church, where there is still a recognition that doctrine grows out of experience and not the other way round. Experience, not doctrine, is the start point in spirituality, as it is in empirical science. By that I take it that Trotter means what I mean: the practice of awareness and the building of a community based on the holistic assumptions that emerge from awareness. Contemplative prayer is not an abstraction: it is the intentional focusing on the dimension of our awareness that points to our participation in the web of reality. Prayer catalyses the building of a loving community. As our research has vividly shown, it is much more than the sentimental avoidance of human solitariness: it is a recognition of the role of our physiology in opening up awareness of transcendence.

Trotter's correspondence theory also reopens the religious option. To recover the fullness of communal living as envisaged by the Peckham Experiment is at the same time to wake up an intuition of transcendence and, Trotter claims, to find once more the living God. The research I have been presenting gives a large hint that for a majority of ordinary people that intuition does indeed lie just under the surface of awareness, suppressed or repressed by the individualism of our culture. Trotter warns that this intuitive discovery will not in itself lead to a restoration of standard theological language, which is likely to remain opaque to a majority of our contemporaries. Does this matter? For the time being Trotter thinks not. We must bear in mind the truth expressed in the apophatic tradition, repeatedly hinted at in the references to 'something there' by the non-churchgoers in Nottingham. No humanly constructed language, including the language of traditional Christian theology, is adequate to the transcendent God.

4. DEVELOPING GENUINE DIALOGUE

Nevertheless, in the end we will need to stammer out our intuitions in community, for it is through shared metaphor that social and spiritual capital are built. For some people, including myself, the corpus of traditional language still holds great and authentic power, genuinely connected to the real experience of life. Naturally I hope that the recovery predicted by Trotter would include an attempt to save that body of metaphor. However, it is worth remembering the wisdom of the Catholic socialist poet, Charles Péguy, who was a great lover of Chartres Cathedral. When World War I broke out, someone expressed a fear that the cathedral might be destroyed during the hostilities. The loss would be unbearable, replied Péguy, but even more unbearable would be the disappearance of the spirit that built the cathedral. To the extent that time-honoured metaphors no longer express the reality of contemporary experience, their very absence will give space for a new dialogue that synthesises the theological, artistic, poetic, mystical and scientific insights of our own time. Alongside the old metaphors, new figures of speech will be born, carrying the vigour and conviction that come from their association with the real human experience of today.

This hope may seem Utopian. How many people outside the religious institutions would be in the slightest degree interested in such matters? The conclusion I draw from the national and international statistics that are now available and the experience of talking with a randomly selected group of non-churchgoers is that *most* people are already deeply interested. They have taught me some surprising truths about the universality and seriousness of the search for ultimate meaning. To put this in terms of traditional Christian language, the Holy Spirit is always present, and in communion with all of created reality. If we are to recover the vivid life of the Spirit that is being offered to us by divine grace, then it must follow that there is much to be learned about the ways of God by listening closely to what so-called 'outsiders' have to say about their search. The problem is not unique to this situation. Spiritual directors (the Irish 'soul friends' is a more appropriate term, because less authoritarian) have the same responsibility, listening to the account of someone's prayer, as do missionaries discerning how God is already manifested in another culture.

The diversity of metaphor created in profusion by humankind

means that very often spiritual discernment requires crossing a divide like that faced by anthropologists trying to understand people from a different culture. Someone who has helpful things to say on this problem is the Japanese student of cross-cultural communication, Muneo Jay Yoshikawa.[46] He is primarily interested in how people from the Far East and Westerners can best communicate with each other, but his proposal of four modes of cross-cultural communication has much to say that relates to spiritual awareness.

- Yoshikawa's *ethnocentric* mode is where I see the people I am meeting, entirely through my own frame of reference. Whatever thoughts and ideas they may have are mere shadows to me and I ignore them. The communication is one-sided and I am deaf to attempts they may make to present their point of view. One might link this mode with an old-fashioned catechism class in which the only purpose is to ensure the successful rote learning of religious beliefs, often without any true understanding. Since I am not even listening to what the other person has to say, there is no way that I can possibly discern the voice of the Spirit in their response.
- The *control* mode is where I scrutinise quite carefully the cultural realities of the people I am encountering. My purpose in taking account of their perspective is, however, entirely manipulative in that I use the information to control and guide them into accepting my point of view.[47] People with a strong missionary desire are often tempted into this mode, which means that, as with the ethnocentric approach, one is deaf and blind to God's Spirit in the other person.
- The *dialectical* mode is where my aim is to achieve a 'fusion' that transcends the differences between myself and those with whom I am in conversation. My assumption in this mode is that if only I was perceptive enough I would see that there are no real differences between us. According to Yoshikawa, in practice such an ideal fusion tends to degenerate into pseudo-dialectic in one of two ways. Either my devotion to the people with whom I am speaking leads me to lose my cultural identity and my beliefs disappear into theirs, or they submerge their identities into mine through politeness or a feeling of being secretly coerced. Apparently liberal religious discussions with non-churchgoers quite often turn into patronising pseudo-dialectic of this kind.

- The *dialogical* mode of encounter requires me to recognise the primacy of the personal relationship between myself and those with whom I am in communication. Whilst it is true that we are separate and independent, we are also simultaneously interdependent and here Yoshikawa is drawing partly from the insights of Martin Buber and also from Buddhist philosophy. There is care to protect the cultural integrity and dignity of both myself and the person with whom I am in conversation. Differences and similarities between us are recognised and respected and the emphasis is on wholeness, mutuality and sensitivity to the immediate dynamics of the meeting, in which God is present. Even when we agree, we each maintain our separate identity and it is important to note that Yoshikawa is referring here to individuality, *not* possessive individualism.

The more we believe that God the Holy Spirit speaks to everyone, the more we are likely to feel that this last dialogical mode is the appropriate way to conduct this kind of conversation. In sensitively trying to enter the space of incomprehension between us, new and unpredictable expressions will emerge, often utilising figures of speech quite remote from the traditional ones. To the extent that they represent an attempt to respond to the Holy Spirit, they are creating new metaphors and new ways of building spiritual solidarity. Speaking from the perspective of my own Christian faith, I believe that during the research in Nottingham I was very often listening to how God the Holy Spirit was already communicating with people who, for one reason or another, keep clear of the institutional church. This listening has been an important part of my own spiritual journey. Our conversations created a mutual reflection on issues of personal depth. When this was at a sufficiently profound level it led inevitably to a questioning of assumptions and a dynamic reconsideration of spiritual issues on my part. Ideally what we discovered together was a non-oppressive, mutually respectful mode of dialogue across the cultural divide that increasingly separates the secular and religious worlds. In the process we found ourselves engaged in the reconstruction of a common spiritual language. I suggest that the development of this dialogue is to discover God's mission to both of us.

5. FACING UP TO THE PROBLEM OF SUFFERING

In August 1961 an American rabbi, Richard Rubenstein, was on a visit to Germany when he had a meeting of which he said, 'No single encounter with another human being ever had so profound an effect upon my religious beliefs'.[48] In the course of a conversation with a prominent German Lutheran pastor in Berlin, the pastor indicated that:

> If I [Rubenstein] truly believed in God as the omnipotent author of the historical drama and in Israel as His Chosen People, I had no choice but to accept [the pastor's] conclusion that Hitler unwittingly acted as God's agent in committing six million Jews to slaughter. I could not believe in such a God, nor could I believe in Israel as the chosen people of God after Auschwitz.[49]

Rubinstein's shocked rejection of the pastor's assertion has shaped his theology ever since. During our research, 'suffering' was the most difficult and unfathomable area for our interviewees. You may also recall that in our national survey, when those interviewed were asked why some people think there is no God, 40 per cent of the respondents referred to the problem of pain; twice as many as the next largest category ('Science has explained away religion').

You don't have to dig very deeply into the average person's life to discover that there are corners of tragedy. Heartbreaking loss and incomprehensible suffering are the lot of most people at some time in their lives, so the issue is not so much how to avoid suffering as how to respond when it comes along. How does spirituality fare in the bad times of someone's life? The answers are contradictory.

George Steiner has pointed to dramatic changes in beliefs about suffering, during the course of European history. In his book *The Death of Tragedy*,[50] published in 1961, Steiner reflected on the classical Greek theatre and its representation of the capriciousness of our fate and our helplessness as playthings of the gods, acted out in great tragedies like *Medea*, the *Oresteia* and *Oedipus Rex*. He contended that with the advent of Judeo-Christian culture in Europe, the classical sense of life as tragic was overcome by the optimism inherent in the belief in a single good God. Judaism and Christianity are just as full of dark events as the Greek tradition. Job's sufferings are archetypal, and the central event in the life of Christ is his crucifixion. But, as Steiner pointed out, in neither case is that the end of the story. Finally God relieves Job of his agonies and he becomes wealthier than ever. Jesus' agony on the Cross is followed by triumphant resurrection. Similarly, for the devout person coping with the terrors of life there is possibly (but see below) the knowledge that behind all the suffering lies a divine plan that will bring good out of evil. The secular successor of this religious tradition, Marxism, similarly rejects the tragic vision because secular salvation is certain to come through the inexorable workings of the dialectic. Steiner reminds us that Anatoly Lunacharsky, commissar for the arts in the first Soviet government, forbade the writing of tragedies as anti-communist.

Such cosmic optimism is put under ever more pressure by the accumulation of disasters. The past hundred years have been full of almost unimaginable catastrophes, ranging from World War I, through to World War II, the Holocaust, the bombing of Hiroshima and Nagasaki, the mass murder of opponents of the regime in the Soviet Union especially during Stalin's leadership, Pol Pot's depradations in Cambodia, and the ethnic cleansing in Kosovo and Rwanda. This is not to mention large-scale natural disasters including the tsunami in 2004 and the destruction of New Orleans in 2005. It is hard to see any of these events as anything other than overwhelming misfortunes and great evils. Even those who are protected from catastrophe by wealth or where they live have these disasters conveyed to them by means of electronic communication in a way that has never happened before in human history.

This is part of modern awareness, and in many of the conversations that we have had there is a strong undertow of frightened avoidance of the realities of life. If at the deepest level there is a conviction that life at depth is pitiless and utterly meaningless, then the optimism of Christianity becomes incredible. Protestations about the goodness of God can sound hollow or unfeeling in such an environment. The impression we gained from our interviewees is that their meetings with functionaries of the church in times of tragedy have often left them wearied by reassuring cliché. The pious assurance that 'All is for the best in this best of all possible worlds' may be arguable philosophically, but the cruel actualities concealed in that statement were exposed long ago by Voltaire in *Candide*.

Forty-three years on from the first publication of *The Death of Tragedy*, with no let-up in the brutalities of life and in a much more secularised world, Steiner returned to his theme to wonder if even the heroics of classical tradegy are possible in a world from which God has departed. He thinks it unlikely:

> The aesthetics of conceptual art, the semen on the bedsheet, the creed of the happening, of Merz or the readymade – reflecting as they do the collapse of agreed values and developing the parodistic genius of Surrealism – are antithetical to high tragedy. Our immediacies are of derision, of black farce, of the multimedia circus.[51]

There are further layers of meaning beneath the mockery. Officially, God is dead but the ghost of God, Beckett's 'bastard who doesn't exist' is still there. Tragedy continues as a kind of dark fairy story and beyond that myth lies something still deeper. It was adverted to most obviously by James during our Nottingham research, but present at least minimally in everyone in the focus groups; the consciousness of 'something there', immediate, meaning-filled and in a different ballpark from the theological chit-chat that James found so empty in university life. During thirty years of research in this field I have come across the story again and again in relation to unbearable suffering; the 'something there' that does not protect us from the pain, but is with us in the midst of it as we struggle through it.

What does the outcome of our research have to say to those concerned with pastoral care or mission in circumstances of personal shipwreck?

- If relationship stays at the level of formal communication then, as Estragon said, there is 'nothing to be done'. Sentimental avoidance along the lines of, 'God has his hidden purposes' is almost certainly the worst approach, though politeness demands (and often gets) a sage nod of agreement on the part of the unfortunate sufferer.
- Silence before the mystery, combined with anger, is the tragic response and is rather more likely to represent contemporary authenticity for a substantial number of people.
- But beyond that – and this is what our informants tell us – in a region of awareness that formally or informally amounts to contemplative prayer, lies something or someone – we are at the apophatic level here – that does not leave us alone.[52] That level of

communication can only take place if the representatives of the institution are themselves deeply prayerful people. Without it, the conversation can never rise above hypocrisy.

BECOMING AWARE THAT RELIGIOUS PEOPLE HAVE ALLIES

I have repeatedly noted that there is a fundamental contradiction between relational consciousness and the possessive individualism that underpins much of the contemporary politico-economic system. Spiritual awareness that is acted upon is subversive of the status quo and, in this respect, committed religious believers share important perspectives with non-believers who are trying to construct a society built on justice and cooperation rather than self-interest. In other words religious people have a large body of allies in the secular world who share their recognition of the social and political importance of relational consciousness, and by implication are not at all hostile to spirituality (remembering the insight that spirituality and religion are not the same thing).

One major reason to expect greatly increased interest in relational consciousness is the equivocal phenomenon of globalisation.[53] Here I want to draw upon the work of my colleague Gordon Lawrence.[54] He has been reflecting for many years on the relation between business management and spirituality, particularly as it relates to the holistic turn taken in scientific understanding during the past thirty years. The so-called 'Anthropic Principle'[55] states that 'the only things that can be known are those compatible with the existence of knowers', implying a profound unity between knower and known. Lawrence reminds us of the simple experiment of raising up one's hand and looking at it. In the act of gazing in this way 'the very cosmos is observing itself'. Next, says Lawrence, turn and glance at some other object, perhaps a tree, and notice that the universe is still observing itself. The same would be true if I had the opportunity to study the distant parts of the universe with the Hubble telescope, or at the other extreme, the nucleus of the atom. Still, one part of the cosmos is observing another. What is being emphasised here is the continuum of which we are constituents. Participation, supremely recognised in relational consciousness, is the true mode of relationship with reality. As I mentioned earlier, the idea of an objective world out there that we observe coldly and scientifically from some

Archimedean point is no longer tenable. The cosmos is one, indivisible whole. From the perspective I have been discussing, to take this deep ecological worldview is to enter the spiritual dimension.

Lawrence points out that these insights have been catalysed by the technological transformation of the world during the past century. For much of human history we have depended on 'work' machines, starting with a piece of wood or bone and culminating in the jet engine. However simple or complex, these machines all share one overriding purpose: to multiply the strength of physical effort. In our own era a new kind of power has become available via electronic machines that hugely multiply the information at our fingertips. The interlocking network of systems is now so obvious that the idea of coercing the environment as an external observer is becoming more and more unrealistic. Currently, says Lawrence, we are entering an age that is governed by a paradigm named by the systems theorist Ludwig von Bertalanffy, 'organismic', and to operate successfully we have to find a symbiosis with it. Lawrence notes that this is particularly clear in Third Wave industries such as:

> … electronics, lasers, optics, genetics, communications, alternative energy, ocean science, space manufacture, ecological engineering, and eco-system agriculture.

The metaphor often used to represent this bewilderingly complex network is to see it as a 'global brain', that is to say, a 'synaptical network' the relevance of which lies in its connections. During the Gulf War and its tragic successor, the Iraq conflict, we began to hear politicians talking of 'linkages', tracing out, as best they could, the causes, effects and consequences of the Middle East conflagration. It quickly became obvious that the linkages proliferate endlessly. As I write they continue to operate throughout world politics, via fear of international terrorism and have material effects in fields like economics and ecology, as well as having more subtle effects on national mood and morale.

As we move more and more into this interlocking network it too requires different social, psychological and spiritual bases for experiencing and thinking through the nature of the relatedness of human beings to their 'cosmos in the mind'. In the new global environment that affects us all, the weakness and, at times, ethical emptiness of a detached hierarchical approach to management becomes clear.[56]

Lawrence points out that more and more managers are experimenting with the idea that their role is to experience as directly as possible in the here-and-now (Donaldson's 'point mode') the business-as-a-series-of-events-in-its-environment. The process is dynamic. In the act of participation, interpretations arise that in turn lead to new insights, enlarged understanding and a fresh cycle of interpretation. As Lawrence describes it this is management by 'revelation' rather than 'salvation'. Whereas in traditional mode the manager steps in with a preconceived plan to 'save' the company, in the ecological mode there is openness, an allowing of oneself to become aware of what is the case and in the process seeing how to go on. The closest analogy to this form of management is with contemplative prayer and it is interesting to note that Lawrence's ideas have partly been drawn from his experience of working as a management consultant with congregations of clergy and religious sisters.

It is difficult to overestimate the importance of the movement described by Lawrence, potentially affecting all levels of management from multinational agencies downwards. It suggests that the individualism that has dominated Western ways of thinking for so long may be losing its economic justification and thus one of the most powerful reasons for its grip on the intellect. The implications for the religious institutions are manifest, primarily because of the rediscovery of the importance of relational consciousness. A changing culture will not leave the churches unscathed. It affects leadership because clergy and others in pastoral roles (as much or more so than managers in other fields) are called to a contemplative rather than a hierarchical way of handling their responsibilities. For all of us it will renew a very longstanding invitation: to repudiate our alienation from our human essence and to rebuild a relationship with the Creator of this beautiful and fragile planet and all who share it. This amounts to the prising open of a cultural valve long choked up, but never quite closed, because at some level people have always known that there is 'something there'.

AN UNSCIENTIFIC POSTSCRIPT

It is better to light a candle than to curse the darkness.
Anonymous

SPIRITUALITY IN UNLIKELY CIRCUMSTANCES

Not long ago I happened to be speaking at a conference in Utrecht, sharing the platform with a man who owed his life to a few seconds of spiritual insight on the part of his enemy. Avraham Soetendorp was born in Amsterdam in 1943 in the most harrowing political circumstances. The German occupying army was in command of the city and the Soetendorp family happened to be Jewish. It is difficult for an outsider to comprehend how much fear there must have been in the hearts of Avraham's parents, desperate to protect their infant son, terrified that he would be killed by the Nazis

One day the inevitable happened. Orders had gone out for the deportation of the Jews and the Soetendorps were waiting with their suitcases packed, ready to be transported to a concentration camp along with the rest of the community in the Juden Viertel. Eventually a Gestapo officer barged into their house and marched into the room where three-month-old Avraham lay. He stared into the cot and muttered 'Pity this is a Jewish child.' With near suicidal courage, perhaps born of despair, Avraham's father reacted, 'I'm happy that he is a Jewish child because whatever happens to him, he will not grow up to be the child of murderers!' The officer struck him violently and shouted at the top of his voice, 'Dirty Jews, you are too filthy to handle; so now we will return tomorrow to take you in.' Then … amazingly … he walked out.

That night the family fled to safety. But why on earth did the Gestapo officer let them escape? Was Avraham spared because he was a good-looking child? Did he awaken a vision of the officer's own children? And did the officer perhaps come to regret his brief moment of weakness? In later years Avraham's father repeated the story, again and again, of how he could see tears in the eyes of the Nazi as he hurried out of the house. Today, like his father before him, Soetendorp is a rabbi. He is a burly, grey-bearded man with an unusually warm and open personality, filled with emotion (as I was) when he spoke of the act of gratuitous mercy that had saved him from almost certain death. His subsequent years have been spent reflecting on what it is that moves a person to transcend hatred and to behave humanely, even courageously, when all the circumstances suggest that savagery is the safe and expected option.

Occasionally Soetendorp has encountered incidents that have shed light on the question. As Chief Rabbi of the Liberal Synagogue in The Hague and President of the European Region of the World Union for Progressive Judaism, he is from time to time involved in political negotiations with representatives of the Palestinian people. One day in the corridor of a Geneva hotel a young Palestinian woman whom he had never met before cornered him and accosted him with angry words: 'You killed my dream!' Controlling his resentment he waited whilst she gave an increasingly heated litany of the injustices and degrading treatment suffered by her people at the hands of the Israelis. As he listened, almost in spite of his annoyance Soetendorp felt a growing sympathy with her and had an instinctive desire to offer her his hand. He hardly dared because she was so furious, but she accepted it. She continued to berate him but her tone moderated. In succeeding years these two from opposite sides of a vicious divide have been involved in further peace negotiations and have become firm friends.[1]

What was the significance of Soetendorp's need to stretch out his hand? He himself felt it was the welling up of an instinctive spirituality. 'I am holding your hand', he seemed to be saying, 'out of some depth more profound than my own conscious resentment, because you and I share a common humanity. That is why, however insufferable the outrages that have happened, I feel impelled to seek reconciliation.'

Of course it is the case that rabbis are supposed to take spirituality seriously. In his professional role Avraham might be expected to be at least potentially alert to his own spirituality, even though at first his dominant feeling was of anger towards his Palestinian accuser. And indeed, after an initial emotional struggle his religious vocation and his here-and-now spiritual insight came into harmony with one another.

But what of a Gestapo officer who contrives to use hate-filled words ('You are too filthy to handle') as a device enabling him to walk away from his murderous duties? Surely he is bound to be in a state of mental discord. The officer who spared the Soetendorps seems to have had a pang of conscience. He was not the only Nazi officer to face this problem.[2]

LEARNING TO CRUSH THE SPIRIT

Evidence presented at the Nuremberg trials of war criminals revealed that the guards in the concentration camps commonly suffered high levels of stress because of the disconcerting way humane feelings would disrupt their peace of mind. This was in spite of their use of a variety of distancing devices, described by Primo Levi who was a prisoner in Auschwitz-Birkenau.[3] To stay detached the guards used tricks of language such as referring to prisoners by numbers rather than names, or repeatedly labelling them as vermin, or subhuman. Physical differences were also useful. Levi interpreted the lack of facilities for hygiene as part of a deliberate plan. Because the prisoners could not keep themselves clean their bodies stank and were therefore disgusting to their tormentors.

In a private speech given to camp personnel in Poznan in 1943, Heinrich Himmler admitted the difficulties when he praised them for pushing their emotions aside:

> Most of you know what it means when 100 corpses are lying side by side, or 500, or 1,000. To have stuck it out and at the same time – apart from exceptions caused by human weakness – to have remained decent men, that has made us hard. This is a page of glory in our history which has never been written and is never to be written.[4]

The hardness that Himmler admired amounts to a refusal of relationship, or in Levinas' language, the supression of the inborn obligation felt by one human being when gazing on the face of the Other.

The strength of this obligation was also apparent in the initial difficulties with members of the *Einsatzgruppen* (death squads set up to kill Jews and members of other out-groups) who felt upset at having to shoot defenceless people, especially women and children. Christopher Browning's study of the experiences of members of one of the *Einsatzgruppen* is an unusually personal and direct study of how, when first confronted with their assignment, the men were traumatised, even though they had been thoroughly indoctrinated to think of Jews as 'vermin'.[5] Most of the members of Reserve Police Battalion 101 were working-class or lower-middle-class men recruited in Hamburg and sent to Eastern Poland. Their initial task in July 1942 was to round up and shoot all the Jews in the village of Józefów, about 60 miles south of the city of Lublin. Major Wilhelm Trapp, their commanding officer, refused to accompany them because he could not bear the sight and was later seen in a highly agitated state, finally bursting into tears 'like a child'. About a dozen men took the opportunity to withdraw from the mission before it started. A much larger group asked to be excused or quietly slipped away when they reached the village and came face to face with the full horror of what they had been ordered to do. These men saw themselves as cowards and weaklings. The remaining members of the *Einsatzgruppe* spent the rest of the day in killing the thousand or so Jews in the village – every last man, woman and child. Eventually this exercise became routine.

THE SPIRIT

There was a lawyer who, to disconcert him, stood up and said to him, 'Master, what must I do to inherit eternal life?' He said to him, 'What is written in the Law? What do you read there?' He replied, 'You must love the Lord your God with all your heart, with all your soul, with all your strength, and with all your mind, and your neighbour as yourself.' 'You have answered right,' said Jesus 'do this and life is yours.' But the man was anxious to justify himself and said to Jesus,

'And who is my neighbour?'

Jesus replied, 'A man was once on his way down from Jerusalem to Jericho and fell into the hands of brigands; they took all he had, beat him and then made off, leaving him half dead. Now a priest happened to be travelling down the same road, but when he saw the man, he passed by on the other side. In the same way a Levite who came to the place saw him, and passed by on the other side. But a Samaritan[6] *traveller who came upon him was moved with compassion when he saw him. He went up and bandaged his wounds, pouring oil and wine on them. He then lifted him onto his own mount, carried him to the inn and looked after him. Next day, he took out two denarii and handed them to the innkeeper. "Look after him," he said "and on my way back I will make good any extra expense you have." Which of these three, do you think, proved himself a neighbour to the man who fell into the brigands' hands?' 'The one who took pity on him,' he replied. Jesus said to him, 'Go, and do the same yourself.'* [7]

Notes

Chapter 1: The Mountains of the Mind

1. Gerard Manley Hopkins, from the poem 'No Worst', reprinted in *The Faber Book of Religious Verse*, ed. Helen Gardner (London: Faber and Faber, 1972).
2. Beckett published the original in French as *En attendant Godot* (Paris: Les Editions de Minuit, 1952/English translation, London: Faber and Faber, 1956). In spite of the hostility of many members of the audiences the play in fact prospered. Not much more than a month after it opened at the Arts Theatre Club, on 12 September 1955, it was moved to a larger theatre, the Criterion.
3. Hugh Kenner draws out the parallels with the knockabout comedy of Laurel and Hardy in *A Reader's Guide to Samuel Beckett* (London: Thames and Hudson, 1973).
4. In 1934 Beckett began a lengthy psychoanalysis with Wilfrid Bion, later to become the founding father of the Tavistock method for the study of group dynamics (see Wilfred Bion, *Experiences in Groups*, London: Tavistock Press, 1961). One of the three major basic assumptions Bion was to identify in the unconscious life of groups is *dependency*, the assumption that if the group waits long enough someone will come and save it. It is tempting to imagine that a similar assumption, at the individual level, emerged from Beckett's unconscious with impressive force during his analysis.
5. The intensity with which Christian ideas pervade all of Beckett's work is brought out particularly well in Mary Bryden's *Samuel Beckett and the Idea of God* (London: Macmillan, 1998).
6. The most extensive and detailed biography of Beckett is James Knowlson's *Damned to Fame: The Life of Samuel Beckett* (London: Bloomsbury, 1996); marginally shorter and more readable is Anthony Cronin's *Samuel Beckett: the Last Modernist* (London: HarperCollins, 1996).
7. This remark was made to Colin Duckworth and reported in his edited book *Samuel Beckett: En attendant Godot* (London: Harrap, 1966).
8. For an aware and insightful discussion of this concordance by a Christian writer, see Richard Harries' essay 'Astride of a grave – Samuel Beckett and Christian hope' in his book of collected essays, *Questioning Belief* (London: SPCK, 1995).
9. Samuel Beckett, *The Unnameable* (English translation, London: John Calder, 1959).
10. See Thomas Cousineau, *Waiting for Godot: Form in Movement* (Boston: Twayne Publishers, 1990), p. 27.
11. John Donne, from 'The first anniversarie: an anatomy of the world' in *The Epithalamions, Anniversaries and Episodes*, ed. W. Milgate (Oxford: Clarendon Press, 1978).

12. Matthew Arnold, from 'Dover Beach', reprinted in *The Penguin Book of English Verse*, ed. John Hayward (London: Penguin Books, 1956), pp. 344–5.
13. W. B. Yeats, from 'The Second Coming', reprinted in *The Penguin Book of English Verse*, p. 407.
14. The late Roy Porter's splendid book, *Enlightenment: Britain and the Creation of the Modern World*, published by Allen Lane (London: Penguin Press, 2000), is a highly readable and scholarly introduction to this complex and controversial concept.
15. The father figure of secularisation theory in Britain was the late Bryan Wilson, formerly Head of the Sociology Department in Oxford University (see his influential text, *Religion in Secular Society*, London: C. A. Watts, 1966). Two important contemporary successors with distinctive perspectives of their own are Callum Brown, Professor of History in the University of Dundee (see *The Death of Christian Britain: Understanding Secularisation 1800–2000*, London and New York: Routledge, 2001) and Steve Bruce, Professor of Sociology at Aberdeen University (see, for example, *God is Dead: Secularization in the West*, Oxford: Blackwell, 2002).
16. Callum Brown identifies the turning point as the 1960s. Before that date, whilst church attendance might not be high, widely held popular assumptions were still recognisably Christian. It was during the 1960s, for example, that it first became impossible to carry on the tradition of singing the hymn 'Abide with me' at the English FA Cup Final. See Brown, *The Death of Christian Britain.*
17. Bruce, *God is Dead*, p. 63.
18. In Peter Brierley, *The Tide is Running Out* (London: Christian Research Association, 2000).
19. Peter Brierley, *Religious Trends No. 1: 1999/2000* (London: Christian Research Association, 1999).
20. Bruce, *God is Dead*, p. 69.
21. Brierley, *Religious Trends.*
22. *Ibid.*
23. *Ibid.*
24. See David Hay and Gordon Heald, 'Religion is good for you', *New Society*, 17 April 1987.
25. They were omitted because we had no money to pay for them. It is an unfortunate economic reality of research in the boundary area between religion and science that such enquiries are usually too scientific for the religious funding bodies and too religious for the agencies supporting science. Consequently (with some notable exceptions) we have often had to make do with not much more than small change.
26. Now called the Religious Experience Research Centre and based in the University of Wales, Lampeter. The current directors are Dr Wendy Dossett, Professor Paul Badham and Professor Xinzhong Yao. The Centre's website is at <http://www.alisterhardytrust.org.uk>
27. Hardy's own preferred classification is presented in his book *The Spiritual Nature of Man* (Oxford: Clarendon Press, 1979). Other attempts include Tim Beardsworth's *A Sense of Presence* (Oxford: Religious Experience

Research Unit, 1977); Meg Maxwell and Verena Tschudin's *Seeing the Invisible* (London: Penguin/Arkana Press, 1990); Geoffrey Ahern's *Spiritual/Religious Experience in Modern Society* (Oxford: Religious Experience Research Centre, 1990).

28. William James, *The Varieties of Religious Experience* (New York: Longman, Green and Co., 1902). There have been many subsequent editions, some with the significant subtitle 'A Study in Human Nature', implying James' belief that the phenomenon has a naturalistic component. The most authoritative edition was published in 1985 as one of the volumes in the Harvard University Press series, *The Works of William James.*
29. Rudolph Otto, *Das Heilige*, translated into English with the title *The Idea of the Holy* (Oxford: Oxford University Press, 1923).
30. Synchronicity has an eminent place in the history of Christianity, for example in the famous story of the conversion of St Augustine near Milan in the year 386. He was standing in a garden in a distressed state when he overheard 'the singsong voice of a child in a nearby house' shouting '*Tolle lege, tolle lege*' ('take it and read, take it and read'). Augustine took the cry to be a direct command from heaven and opening at random a book containing St Paul's Epistles, read the words, 'Not in revelling and drunkenness, not in lust and wantonness, not in quarrels and rivalries. Rather, arm yourselves with the Lord Jesus Christ; spend no more thought on nature and nature's appetites.' [Described in Book VIII xii (29) of Augustine's *Confessions*]. In the same paragraph, Augustine himself refers to a precedent for his experience; the conversion of St Anthony in a church in Alexandria when he heard the words from St Matthew's Gospel, 'Go, sell all you have, give to the poor, and you shall have treasure in heaven; and come, follow me' (Matt. 19:21).
31. Edward Robinson made a moving study of the religious experience of childhood, drawing upon such accounts in the RERV archive. See his book, *The Original Vision* (New York: Seabury Press, 1983).
32. See note 24.
33. The point where Wordsworth begins to speak more overtly of his experience of transcendence is from line 415 onwards, from *The Prelude* Book I: Childhood and school-time (Oxford: Oxford University Press, 1977):

    I left my Bark,
    And, through the meadows homeward went, with grave
    And serious thoughts; and after I had seen
    That spectacle, for many days, my brain
    Work'd with a dim and undetermin'd sense
    Of unknown modes of being; in my thoughts
    There was a darkness, call it solitude,
    Or blank desertion, no familiar shapes
    Of hourly objects, images of trees,
    Of sea or sky, no colours of green fields;
    But huge and mighty Forms that do not live
    Like living men mov'd slowly through the mind
    By day and were the trouble of my dreams.

34. Director of the Religious Experience Research Unit from 1976 to 1985.
35. Quoted in *The Original Vision*, p. 35.

36. See for example W. T. Stace's important book *Mysticism and Philosophy* (London: Macmillan, 1960).
37. In David Tacey, *The Spirituality Revolution: the Emergence of Contemporary Spirituality* (Sydney: HarperCollins, 2003), p. 14.
38. Ibid.
39. See Grave Davie, *Religion in Britain since 1945: Believing without Belonging*, with an introduction by David Martin (Oxford: Blackwell, 1994).
40. Personal communication from Yves Lambert. See his complete paper, 'A turning point in religious evolution in Europe', *Journal of Contemporary Religion*, 19:1 (2004), pp. 29–45.
41. See B. J. Zinnbauer *et al.*, 'Religion and spirituality: unfuzzying the fuzzy', *Journal for the Scientific Study of Religion* 76:4 (1997), pp. 549–64.

Chapter 2: 'Unfuzzying the Fuzzy': Working Towards a Biology of the Spirit

1. From the title of Zinnbauer's important paper. See ch. 1, note 38.
2. Luke 4:18–19.
3. From Eugene V. Debs 'Statement to the Court Upon Being Convicted of Violating the Sedition Act' because of his opposition to the entry of the United States into World War I.
4. B. V. Maturin, *Laws of the Spiritual Life* (London: Longman, Green and Co., 1909). Basil Maturin was an Irishman of Huguenot stock, born in 1847. He entered the Anglican Cowley Fathers in Oxford and eventually became a Roman Catholic. In 1913 he was appointed Catholic Chaplain at Oxford University.
5. *Ibid.*, p. 122.
6. Maturin himself had already been a casualty, being drowned when the *Lusitania* was sunk by a German U-boat in 1915. See Maisie Ward, *Father Maturin: A Memoir* (London: Longman, Green and Co, 1920).
7. See Ray Ginger, *Eugene Debs: A Biography* (New York: Simon and Schuster, 1962).
8. World Spirituality series, general editor: Ewart Cousins (New York: Crossroad London: SCM Press)
9. *Spirituality and the Secular Quest* (World Spirituality series), edited by Peter H. Van Ness (New York: Crossroad/London: SCM Press, 1996).
10. *Guardian Weekend*, 28 February 2004, pp. 41–55.
11. In Marx's introduction to his 'Contribution to the critique of Hegel's philosophy of right' (*Deutsch-Französische Jahrbücher*, February 1844); republished in K. Marx and F. Engels, *On Religion* (Moscow: Progress Publishers, 1972), p. 38.
12. See Nancy Bancroft, 'Spirituality in Marxism: a Communist view', in Benjamin B. Page (ed.) *Marxism and Spirituality: An International Anthology* (Westport, Connecticut: Bergin and Garvey, 1993), p. 84.
13. *Ibid.*, pp. 85–6.
14. Reproduced in J. M. Cohen's introduction to his translation of Pascal's *Pensées* (London: Penguin Classics, 1961).
15. See J. M. Cohen and J-F. Phipps, *The Common Experience* (London: Rider and Company, 1979), p. 137.
16. An instance of Beckett's compassionate feelings and respect for religious

believers is recorded in a letter written by a colleague who worked with him in a military hospital near Paris after World War II:

> Sam [Beckett] has an assistant storekeeper here named Tommy Dunne, a very decent little Dublin chap. Sam is … in every way a most likeable chap, aged about 38–40, no religious persuasion, I should say a free thinker – but he pounced on a little rosary beads (sic) which was on a stall in Notre Dame to bring back as a little present to Tommy D.

Referred to in James Knowlson, *Damned to Fame: the Life of Samuel Beckett* (London: Bloomsbury, 1997), p. 347.

17. Vítězslav Gardavsky, *God Is Not Yet Dead* (London: Penguin Books, 1973), pp. 83–4.
18. See Mircea Eliade (ed.), *The Encyclopedia of Religion*, Vol. 6 (New York: Macmillan, 1986), pp. 313–18.
19. Translated from the fourth-century Greek original by Gerard Moultrie, 1864. <http://www.cyberhymnal.org/htm/l/e/letallmf.htm>.
20. From the poem *The Elixir*, in George Herbert, *The Complete English Poems* (London: Penguin Books, 2004).
21. Ludwig Feuerbach, *The Essence of Christianity*, tr. George Eliot with an introduction by Karl Barth (New York: Harper Torchbooks, 1957), p. 73.
22. It is worth adding that whilst Paley is best known for his defence of the argument from design, his own religious faith was based much more on personal experience, hence quite closely linked to the natural theology advocated by Hardy. See D. L. le Mahieu, *The Mind of William Paley* (Nebraska: University of Nebraska Press, 1976).
23. See Richard Dawkins, *The Blind Watchmaker* (London: Longman Scientific and Technical, 1986).
24. An extended account of Hardy's association with Aberdeen University can be found in my essay, 'A biologist of God: Alister Hardy in Aberdeen', *Aberdeen University Review* LX 3, no. 211 (2004), pp. 208–21.
25. Or so says the Scottish Commonsense School of Philosophy, the leading light of which was the eighteenth-century Aberdeen philosopher Thomas Reid. See *Thomas Reid's Inquiry and Essays*, edited by Ronald E. Beanblossom and Keith Lehrer (Indianapolis: Hackett Publishing Company Inc., 1983).
26. Alister Hardy, *The Living Stream: A Restatement of Evolution Theory and its Relation to the Spirit of Man* (London: Collins, 1965).
27. Published by Stanford University Press.
28. Alister Hardy, *The Divine Flame* (London: Collins, 1966).
29. Discussed by David Wulff in his book *Psychology of religion: Classic and contemporary*, 2nd edn (New York: Wiley and Sons, 1997).
30. See Eduard von Hartmann's book, *Das Religiöse Bewussstsein der Menschheit im Stufengang seiner Entwicklung* (Berlin: Carl Buncker, 1882).
31. See Wolfgang Köhler, *The Mentality of Apes*, tr. Ella Winter, 2nd revised edn (London: Routledge and Kegan Paul, 1921).
32. See J. Malan, 'The possible origin of religion as a conditioned reflex', *American Mercury* 25 (1932), pp. 314–17.
33. Published by Jonathon Cape in London.

34. See Ralph Solecki, *Shanidar: The First Flower People* (New York: Knopf, 1971), also Arlette Leroi-Gourhan, 'The flowers found with Shanidar IV, a Neanderthal burial in Iraq', *Science* 190 (1975), 562–4.
35. See James Shreeve, *The Neanderthal Enigma: Solving the Mystery of Modern Human Origins* (New York: William Morrow and Company Inc., 1995).
36. For a discussion of the role of religion in coping, see Kenneth Pargament's book *The Psychology of Religion and Coping: Theory, Research, Practice* (New York/London: Guilford Press, 1997).
37. Robert Marrett, from *Head, Heart and Hands in Human Evolution* (London: Hutchinson, 1935).
38. Émile Durkheim, *The Elementary Forms of the Religious Life*, tr. J. W. Swain (London: George Allen and Unwin, 1915), p. 416.
39. *Ibid.*, pp. 423–4.
40. At the beginning of the nineteenth century the German Pietist theologian Friedrich Schleiermacher had worked out an entire systematic theology on the basis of what he called 'the feeling of absolute dependence'. At the very end of the same century Edwin Starbuck, a student in the Harvard Divinity School, made the first systematic attempt to explore contemporary religious experience using scientific methods. Since he was working in New England, a region uniquely dominated by its history of Puritan colonisation, religious experience effectively meant for him the experience of conversion. Soon afterwards William James' masterly Gifford Lectures in Edinburgh University were to depend heavily on Starbuck's data for their empirical content, though his claims for universality meant he went far beyond the Puritan conversion experience. Meanwhile in Europe Ernst Troeltsch, Rudolf Otto and, later, Joachim Wach, as leading students of the experiential dimension of religion, found an intellectual source for their preoccupations in the work of Schleiermacher (see the bibliography for works by these scholars).
41. See Anselm Stolz, *The Doctrine of Spiritual Perfection* (St Louis: B. Herder Book Co., 1938).
42. See Auguste Poulain, *The Graces of Interior Prayer* (London: Kegan Paul, Trench, Trubner and Co., 1912).
43. For example, see Joseph Maréchal, *Studies in the Psychology of the Mystics* (London: Burns, Oates and Washbourne, 1927).
44. I discuss the rivalry of these two men in Chapter 7. See also my article 'Psychologists interpreting conversion: two American forerunners of the hermeneutics of suspicion', *History of the Human Sciences* 12:1, (1999), pp. 55–72.
45. Hardy was very interested in parapsychology and was for a time President of the British Society for Psychical Research. He also cooperated with Arthur Koestler and Robert Harvie in writing a book about it, *The Challenge of Chance* (London: Random House, 1973).
46. Published in Lodon by Chatto & Windus.
47. See Steven T. Katz (ed.), *Mysticism and Philosophical Analysis* (New York: Oxford University Press, 1978).
48. Much of this argument is taken from David Hay, 'Asking questions about religious experience', *Religion* 18 (1988), pp. 217–29.

49. See Geertz, 'The way we think now: toward an ethnography of modern thought', in *Local Knowledge: Further Essays in Interpretive Anthropology* (New York: Basic Books, 1983).
50. See for example, the Jesuit William Johnson's book *Silent Music: The Science of Meditation* (London: Collins, 1974); J. K. Kadowaki, SJ, *Zen and the Bible: A Priest's Experience* (London: Routledge and Kegan Paul, 1980); the Cistercian Thomas Merton's *Zen and the Birds of Appetite* (New York: New Directions, 1968); and the Benedictine monk, Aelred Graham's *Conversations: Christian and Buddhist* (New York: Harcourt Brace Jovanovich, 1968).
51. I am grateful to my colleague Philip Endean SJ for suggesting the use of the term 'common context'.
52. There is a more complete review of much of this work in my paper, '"The biology of God": what is the current status of Hardy's hypothesis?', *International Journal for the Psychology of Religion* 4:1 (1994), pp. 1–23.
53. e.g. Charles Glock and Rodney Stark, *Religion and Society in Tension* (Chicago: Rand McNally, 1965); Andrew Greeley, *The Sociology of the Paranormal: A Reconnaissance* (Sage Research Papers in the Social Sciences: Studies in Religion and Ethnicity Series no. 90–023, Beverley Hills/London: Sage Publications, 1975); Robert Wuthnow, *The Consciousness Reformation* (Berkeley: University of California Press, 1976); Morgan Poll, 'Unpublished poll of reports of religious experience in Australia', 1983; Gallup Poll, USA, '4 in 10 Americans have had unusual spiritual experiences' (Press release,1985); Sabino Acquaviva, *Eros morte ed esperienza religiosa, (Eros, Death and Religious Experience)* (Bari: Laterza, 1991).
54. See Hay, '"The biology of God": what is the current status of Hardy's hypothesis?' op. cit.
55. The defensive functions of distorted religious experience have been discussed by the priest psychoanalyst André Godin in his book *The Psychological Dynamics of Religious Experience* (Birmingham, Alabama: Religious Education Press, 1985).
56. Friedrich Engels, 'Letter from Wuppertal' under the pseudonym Friedrich Oswald, reprinted in Marx and Engels, *Collected Works* (London: Lawrence and Wishart, 1975), vol. II, p. 578.
57. Cf. A. P. Donajgrodski (ed.), *Social Control in Nineteenth-century Britain* (London: Croom Helm, 1977).
58. See Freud's paper 'A religious experience', first published in 1928 and reprinted in the *Standard Edition of the Works of Sigmund Freud* (London: Hogarth Press, 1959), vol. XXI, p. 169.
59. This more extreme claim was made by Theodor Reik in *From Thirty Years with Freud* (New York: Farrar and Rinehart, 1941). It tallies with some major psychiatric diagnostic instruments that advise psychiatrists to treat certain kinds of alleged religious experience as symptomatic of mental illness. I discuss this in Chapter 4.
60. The scale used was the Balanced Affect Scale, developed by Norman Bradburn at the University of Chicago. See his book, *The Structure of Psychological Wellbeing* (Chicago: Aldine Press, 1969).
61. See Greeley, *The Sociology of the Paranormal: A Reconnaissance.*

62. See David Hay and Ann Morisy, 'Reports of ecstatic, paranormal or religious experience in Great Britain and the United States: a comparison of trends', *Journal for the Scientific Study of Religion* 17:3 (1978), pp. 255–68.
63. Sigmund Freud, *The Future of an Illusion* (London: Hogarth Press, 1928).
64. Godin, *The Psychological Dynamics of Religious Experience.* Op. cit.
65. The classic work on this subject is still *The Authoritarian Personality*, by Theodor Adorno, Else Frenkel-Brunswik, Daniel J. Levinson and R. Nevitt Sanford (New York: Harper and Row, 1950).
66. As Gordon Allport put it many years ago in *The Individual and His Religion* (New York: Macmillan, 1962), temperament must play a part, with extraverts preferring the stimulation of a crowd. Nevertheless such people were a small minority in the British sample.

Chapter 3: The Individuality of the Spirit

1. See David Hay, 'Memories of a Calvinist Childhood', in W. Gordon Lawrence (ed.) *Roots in a Northern Landscape: Celebrations of Childhood in the North East of Scotland* (Edinburgh: Scottish Cultural Press, 1996), pp. 48–61.
2. See H. Schmid, *The Doctrinal Theology of the Evangelical Lutheran Church*, tr. Charles A. Hay and Henry E. Jacobs (Minneapolis: Augsburg Publishing House, 1961 [1876]).
3. *The Spiritual Life: A Treatise on Ascetical and Mystical Theology* by the Very Reverend Adolphe Tanquerey SS, DD, tr. Herman Branderis (Tournai/Paris/Rome/New York: Desclee and Co., 1930). Amongst literally hundreds of similar manuals, two other well-known examples are Auguste Poulain's *The Graces of Interior Prayer*, tr. Leonora L. Yorke Smith (London: Kegan Paul, Trench, Trubner and Co., 1928); and Dom Vital Lehodey's *The Ways of Mental Prayer*, tr. by a monk of Mount Melleray (Dublin: M. H. Gill and Son Ltd.,1924).
4. *The Spiritual Life* is still in print and available from internet booksellers, though hardly a bestseller. When I looked it up in the Amazon online publisher's list recently, it was ranked 592,338th in popularity. Nevertheless, for some people it retains a down-to-earth practicality. One reviewer in Amazon gives it five stars and says:

   > For the past 20 years I have used a copy I bought second-hand and it is the book I refer to more than any other. It is clearly laid out – every paragraph is numbered for easy reference – and is full of practical, sensible advice.

5. See Table 2 on p. 265.
6. An intuition supported by the work of Callum Brown at Strathclyde University. See his book *The Death of Christian Britain*, op. cit. Chaptr 1.
7. The focus groups were led by Gordon Heald, David Hay and Kate Hunt. With the permission of the participants, each group was both audio and video recorded. It was felt that the benefits of being able to review the body language of the participants at a later stage outweighed the possibly inhibiting presence of the video camera. Individual research discussions were managed by David Hay and Kate Hunt (who did the majority of the fieldwork). A list of the themes covered in the discussions included:

Table 2: *Structure of the Focus Groups*

	Gender	Self-Identification	Age Range	Church Background
Focus Group 1	4 women 3 men	5 spiritual 2 religious	25–35	2 C of E 1 RC 1 Methodist 1 Pentecostal 2 none
Focus Group 2	5 women 3 men	4 spiritual 4 religious	40–55	2 C of E 2 RC 1 Nonconf. 3 none
Focus Group 3	8 women	3 spiritual 5 religious	27–60	3 C of E 2 RC 2 Baptist 1 none
Focus Group 4	8 men	5 spiritual 3 religious	24–45	2 C of E 1 Methodist 1 Evangelical 4 none

- Background information on childhood – any experience of organised religion?
- Critical incidents in your life
- Experiences of transcendence
- Meaning and purpose in life
- Attitudes towards the family and the wider community
- What does it mean to be 'spiritual'?
- Do you talk about these issues with other people?
- Issues around the institutional Church and Christian beliefs (including the Bible)
- Belief in God
- Issues around death and the possibility of an after-life

8. Diarmuid Ó Murchú, *Reclaiming Spirituality* (Dublin: Gill and Macmillan, 1997), p. 61.
9. The term comes from the American sociologist of religion Robert Wuthnow.

Chapter 4: Shared Aspects of the Spiritual Quest

1. Ireland is no longer immune from institutional decline, so Mary's reference to her home country may soon be out of date.

2. See American Psychiatric Association, *Diagnostic and Statistical Manual of Mental Disorders*, 3rd revised edn, DSM-3 (Washington DC: American Psychiatric Association, 1987). At the time of writing there have been two subsequent editions, DSM-4 and DSM-5.
3. See DSM-3, p. 188.
4. Arthur Hugh Clough, from 'There is no God the wicked sayeth' in *The Poems of Arthur Hugh Clough*, edited by F. L. Mulhauser, 2nd edn (Oxford: Clarendon Press, 1974).

Chapter 5: Theorising about the Spirit

1. W. B. Yeats, from 'Song of an Old Man' in *The Collected Poems of W. B. Yeats* (Ware: Wordsworth Editions, 2000).
2. Matthew Arnold, from 'Dover Beach' in Matthew Arnold, *Poems*, introduction by Kenneth Allott, new edn (London: Dent, 1965).
3. See references to Zinnbauer *et al.*, Hay and Hunt, and Lambert in ch. 1.
4. Luke 18:9–14.
5. See Voltaire, *Candide*, edited with introduction, notes and bibliography by Haydn Mason (London: Bristol Classical Press, 1995).
6. The criticism came from Columba Marmion, Abbot of the Benedictine abbey of Maredsous in Belgium. Marmion's words, as quoted by Dom John Chapman who was a monk at Maredsous were, 'St John of the Cross is like a sponge full of Christianity. You can squeeze it all out, and the full mystical theory remains.' Chapman adds that, 'Consequently, for fifteen years or so, I hated St John of the Cross and called him a Buddhist.' See Dom John Chapman OSB, *Spiritual Letters* (London and Sydney: Sheed and Ward, 1935), p. 269.
7. St John of the Cross, from Book I, Chapter 13 of *The Ascent of Mount Carmel.* See *The Collected Works of St John of the Cross*, tr. Kieran Kavanaugh OCD and Otilio Rodriguez OCD (Washington DC: Institute of Carmelite Studies, 1973), pp. 103–4.
8. Reported in Vygotsky, *Thought and Language*, 2nd edn (Cambridge, Massachusetts: MIT Press, 1986).
9. Even so-called discovery methods in science education are mostly exercises in 'discovering' what the teacher has decided you will discover.

Chapter 6: Primordial Spirituality

1. See Karl Rahner, 'The experience of God today' in *Theological Investigations XI*, tr. David Bourke, (London: Darton, Longman and Todd, 1974), ch. 6, pp. 149–65.
2. Alison Gopnik, *Guardian G2*, 7 January 2005, p. 7.
3. I am in no doubt about the importance of social construction in my own case. See my essay, 'Memories of a Calvinist childhood', in W. Gordon Lawrence (ed.), *Roots in a Northern Landscape.* Op. cit.
4. One of the best expositions of the social origins of the self is Charles Taylor's *Sources of the Self: The Making of the Modern Identity* (Cambridge: Cambridge University Press, 1989).
5. Richard Dawkins, in *Free Inquiry Magazine* 24 (5), 2004. Posted on the Internet at <www.santafe.edu/~shalizi/Dawkins/viruses-of-the-mind.html>.

6. Aaron Lynch specifically uses the metaphor of disease to interpret the spread of religious belief in his book *Thought Contagion: the New Science of Memes* (New York: Basic Books, 1996).
7. For example the Yale scholar George Lindbeck in his book *The Nature of Doctrine: Religion and Theology in a Post-Liberal Age* (Philadelphia: Westminster Press, 1984).
8. See Charles J. Lumsden and Edward O. Wilson, *Promethean Fire: Reflections on the Origin of Mind* (Cambridge, Massachussetts: Harvard University Press, 1983), p. 54.
9. See Leslie A. White, 'Individuality and Individualism: a culturological interpretation', *Texas Quarterly* 6:2 (1963), pp. 111–27. White's attitude to genetics is reminiscent of Stalin's favourite botanist, Trofim Lysenko, who adopted a strong version of Lamarckism in his doctrines of plant breeding and in the process destroyed much of Soviet biology. See David Joravsky, *The Lysenko Affair* (Cambridge, Massachussetts: Harvard University Press, 1970, republished by the University of Chicago Press, 1986).
10. Lumsden and Wilson, *Promethean Fire*, p. 54.
11. See C. D. Darlington, *The Evolution of Man and Society* (London: Allen and Unwin, 1969).
12. Lumsden and Wilson, *Promethian Fire*, pp. 57, 60.
13. Richard Dawkins, in *The Selfish Gene* (Oxford: Oxford University Press, 1976).
14. For a readable critique of meme theory as deployed against religion, see Alister McGrath, *Dawkins' God: Genes, Memes and the Meaning of Life* (Oxford: Blackwell Publishing, 2004).
15. The way memetics is often presented makes it sound like a new kind of Cartesian dualism, at a time when the philosophical consensus is strongly against dualism. Introductions to memetics include Susan Blackmore's *The Meme Machine* (Oxford: Oxford University Press, 1999), and Robert Aunger's edited volume *Darwinising Culture: the Status of Memetics as a Science* (Oxford: Oxford University Press, 2001).
16. St Augustine, *Confessions*, available in numerous editions, including a modern translation by Henry Chadwick (London: Penguin Books, 1991).
17. Kalevi Tamminen, *Religious Development in Childhood and Youth* (Helsinki: Suomalainen Tiedeakatemia, 1991).
18. Auguste Comte, *The Positive Philosophy of Auguste Comte*, tr. Harriet Martineau, vol. 1 (London, Batoche Books, 2000/1853), p. 25.
19. Full details of our discussion and the process of identification can be found in Rebecca Nye and David Hay, 'Identifying children's spirituality: how do you start without a starting point?', *British Journal of Religious Education* 18:3 (1996).
20. See Nyanaponika A. Thera, *The Heart of Buddhist Meditation* (London: Rider, 1969).
21. Many years ago I learned *vipassana* under the tuition of a Buddhist monk, who encouraged awareness at all times, whether eating, drinking, going to the lavatory or even going to sleep. At times this felt uproariously funny, in which case the deadpan instruction was to be aware 'there is laughter' as a continuation of the meditative state.

22. J-P. de Caussade, *Self Abandonment to Divine Providence*, tr. Algar Thorold (London: Burns and Oates, 1933).
23. See Margaret Donaldson, *Human Minds* (London: Allen Lane, 1992).
24. The psychologist Alison Gopnik, talking about her beliefs in the *Guardian G2* section, 7 January 2005. See also Alison Gopnik, Andrew Meltzoff and Patricia Kuhl, *The Scientist in the Crib: Minds, Brains and How Children Learn* (New York: HarperCollins, 2001).
25. Martin Heidegger, *Being and Time*, tr. John Macquarrie and Edward Robinson (Oxford: Blackwell, 1962). First published in 1927 as *Sein und Zeit*, this book is notoriously difficult, but Hubert Dreyfus has produced a very helpful commentary, *Being-in-the-World* (Cambridge, Massachusetts: MIT Press, 1991).
26. I am grateful to Philip Green Educational Ltd for allowing us to use the pictures and supplying them free of charge.
27. Rebecca gives a detailed and subtly nuanced account of her work in our book *The Spirit of the Child* (London: HarperCollins, 1998). See chs 6 and 7.
28. For more detail, see ch. 7 of Hay and Nye, *The Spirit of the Child*.
29. See the *User's Guide for QSR.NUD*IST* (Sage software: SCOLARI, 1996).
30. Hay and Nye, *The Spirit of the Child*, p. 113.
31. *Bhagavadgita* 6:10.
32. Matthew 6:6.
33. The terms 'holism' and 'holistic' were introduced by the South African politician and thinker Jan Christiaan Smuts in his book *Holism and Evolution* (London: Macmillan, 1926). Dated in some ways, the book nevertheless expresses ideas that underpin modern Green politics.
34. I am grateful to my colleague Adrian Thatcher for catalysing my thoughts on this matter. See David Hay and John Hammond, ' "When you pray, go to your private room": a reply to Adrian Thatcher', *British Journal of Religious Education* 14:3 (1992).
35. See Olga Moratos, *The Origin and Development of Imitation in the First Six Months of Life*, PhD thesis submitted to the University of Geneva in 1973.
36. See the many writings of Colwyn Trevarthen on infant intersubjectivity, for example, 'The self born in intersubjectivity, an infant communicating', in U. Neisser (ed.), *The Perceived Self: Ecological and Interpersonal Sources of Self Knowledge* (New York: Cambridge University Press, 1993); also, 'The neurobiology of early communication: intersubjective regulation in human brain development', in A. F. Kalverboer and A. Grimsbergen (eds) *Handbook of Brain and Behavior in Human Development* (Dordrecht: Kluwer Academic Publications, 2001); also, on early intellect, Gopnik *et al.*, *The Scientist in the Crib*.
37. See Emese Nagy and Peter Molnar, 'Homo imitans or Homo provocans? Human imprinting model of neonatal imitation', *Infant Behavior and Development* 27 (2004), pp. 54–63. At the time of writing, Dr Nagy is a member of staff at the University of Dundee.
38. I first saw this videotape at a conference in the University of Aberdeen in 1996, where it was shown by Professor Colwyn Trevarthen of Edinburgh University.
39. See Lynn Murray and Liz Andrews, *The Social Baby* (London: CP Publishing, 2005).

Chapter 7: Psychologists Start Arguing about Spirituality

1. From 'Religion as a cultural system' in Clifford Geertz, *The Interpretation of Cultures* (New York: Basic Books Inc., 1973), p. 123.
2. See E. D. Starbuck, 'Religion's use of me', in Vergilius Ferm (ed.), *Religion in Transition* (London: George Allen and Unwin, 1937), p. 226.
3. A similar but much less elaborate investigation by James Leuba had been published in the *American Journal of Psychology*, vol. VII (1896), p. 322. An edited version was republished as *The Psychology of Religious Mysticism* (London: Kegan Paul, Trench, Trubner and Co., 1925).
4. Starbuck, op. cit.
5. Bornkamp attended the Divinity School for just over one year. Ordained as an Episcopalian priest in Boston in 1897, he served parishes in Duxbury, Boston, and Winona, Minnesota.
6. Starbuck, op. cit.
7. Higginson commanded the first black regiment in the United States army in 1862 and corresponded with Darwin.
8. One interesting sidelight to this story concerns Estlin Carpenter, later to become Principal of Manchester College in Oxford. At the time when Starbuck was doing his research Carpenter was a visiting Chaplain at Harvard and took a packet of questionnaires back to Oxford for use amongst his students. So far as I am aware these are the only questionnaires used outside Harvard. By a remarkable coincidence, 75 years later Sir Alister Hardy set up the Religious Experience Research Unit in Manchester College with closely similar objectives to those of Starbuck.
9. E. D. Starbuck, *The Psychology of Religion* (London: Walter Scott, Paternoster Square, 1899). See especially ch. XII.
10. This view pervades William James' great two-volume work *The Principles of Psychology* (New York: Henry Holt, 1890). James' sources for the idea were the German psychologists Wilhelm Wundt and Gustav Fechner. The notion of psychophysical parallelism is of course a particular psychological response to a much more longstanding philosophical problem i.e. the relation between mind and body.
11. *Ibid.*, p. ix.
12. I have discussed this in more detail elsewhere. See David Hay with Rebecca Nye, *The Spirit of the Child*, revised edn (London/New York: Jessica Kingsley Publishing, 2006). See especially Chapter 2.
13. See Roy Porter's *Enlightenment: Britain and the Creation of the Modern World*, op. cit.Starbuck, op. cit. Chapter 1.
14. For a clear summary of this important distinction see 'The hermeneutics of suspicion and retrieval: Paul Ricœur's hermeneutical theory', ch. 6 of Anthony Thiselton's book *New Horizons in Hermeneutics* (London: HarperCollins, 1992).
15. Published as *The Varieties of Religious Experience* (New York: Longman, Green and Co., 1902).
16. The eminent Canadian philosopher Charles Taylor has produced an illuminating study, both critical and laudatory, of the *Varieties*. See *Varieties of Religion Today: William James Revisited* (Cambridge: Massachusetts: Harvard University Press, 2002).
17. Jonathan Edwards, *The Religious Affections* (republished Edinburgh:

Banner of Truth Trust, 1986).

18. Quoted by Perry Miller, 'Jonathan Edwards on the Sense of the Heart', *Harvard Theological Review* 41 (1948).
19. *Ibid.*, p. 124.
20. See *The Letters of William James,* ed. Henry James (Boston: Atlantic Monthly Press, 1920), vol. II, p. 127.
21. In his 1892 review of James' *Principles of Psychology*, Myers made the following comment:

    > I suggest then, that the stream of consciousness in which we habitually live is not the only consciousness which exists in connection with our organism. Our habitual or empirical consciousness may consist of a mere selection from a multitude of thoughts and sensations of which some at least are equally [capable of becoming?] conscious with those we empirically know. I accord no primacy to my ordinary waking self, except that among my potential selves this one has shown itself the fittest to meet the needs of the common life.

    See F. W. H. Myers, 'The Principles of Psychology', *Proceedings of the Society for Psychical Research,* vol. 7 (1892), p. 301.
22. James, *Varieties*, p. 487.
23. *Ibid.*, p. 490.
24. See Cushing Strout, 'The pluralistic identity of William James', *American Quarterly* 23:2 (1971), p. 135.
25. The myth of the detached scientist isolated from personal bias began to lose credibility during the 1960s. One key figure in the process of deconstruction is Hans-Georg Gadamer. His master work *Warheit und Methode* appeared first in 1960 and was translated into English in 1975. The second English edition was entitled *Truth and Method* (London: Sheed and Ward, 1989). My own dogmatic slumbers were first rudely disturbed in 1962 by the publication of Thomas Kuhn's book *The Structure of Scientific Revolutions* (Chicago: University of Chicago Press, 1962).
26. See J. H. Leuba, 'Professor William James' interpretation of religious experience', *International Journal of Ethics* 14 (1903–4), pp. 322–9.
27. Edwin married Anna Diller, a fellow student in the Harvard Divinity School. Anna was blind and Edwin must have admired her greatly for after his marriage he inserted her surname into the middle of his own name.
28. Starbuck, 'Religion's use of me', p. 215.
29. *Ibid.*
30. Published in London in 1873 by Longmans, Green and Co.
31. Starbuck was frequently criticised for the weaknesses of the questionnaire method. See for example, G. M. Stratton, *Psychology of the Religious Life* (London: George Allen and Unwin, 1911); George Coe, *The Psychology of Religion* (Chicago: University of Chicago Press, 1916), pp. 44–7. In spite of these criticisms, in practice Starbuck was very cautious in most of his interpretations of his data.
32. Here Starbuck is anticipating the ideas of J. B. Rotter. See *Social Learning and Clinical Psychology* (Englewood Cliffs, NJ: Prentice Hall, 1954).
33. See George Coe, *The Spiritual Life* (New York: Fleming H. Revell Co., 1900).

34. Starbuck, *The Psychology of Religion*, p. 77.
35. *Ibid.*, p. 78.
36. See W. B. Carpenter, *Mental Physiology* (New York: D. Appleton and Co., (1881), p. 344. I am informed that Carpenter had developed the notion of 'unconscious cerebration' well before this date, in the fourth edition of his *Human Physiology*, published in 1852 (personal communication).
37. Starbuck, *The Psychology of Religion*, p. 111.
38. *Ibid.*, p. 148.
39. Gustave Le Bon, *The Crowd* (New York: Dover Publications, 2002).
40. Starbuck, *The Psychology of Religion*, p. 168.
41. Quoted by Starbuck in 'Religion's use of me', Sidis was an implacable opponent of stances such as that of Starbuck. See for example his remarks on religion in *The Psychology of Suggestion* (New York: D. Appleton, 1903), especially chs 17, 18 and 23.
42. The film actress Katharine Hepburn remembered his kindness to her when she was a psychology student at Bryn Mawr.
43. Taken from, 'The making of a psychologist of religion', in Vergilius Ferm (ed.), *Religion in Transition*, p. 178.
44. *Ibid.*, p. 180.
45. In spite of Comte, the second of whose three stages of growth towards the scientific attitude was the 'metaphysical'.
46. J. H. Leuba, 'A study in the psychology of religious phenomena', *American Journal of Psychology*, vol. 8 (1896), p. 322. Also published separately as *Studies in the Psychology of Religious Phenomena* (Worcester, Massachusetts: J. H. Orphe, 1896).
47. J. H. Leuba, *The Psychology of Religious Mysticism* (London: Kegan Paul, Trench, Trubner and Co., 1925).
48. *Ibid.*, p. 157.
49. *Ibid.*, p. 140.
50. *Ibid.*, p. 200.
51. For a contemporary application of attribution theory to religious experience, see Wayne Proudfoot and Philip Shaver, 'Attribution theory and the psychology of religion', *Journal for the Scientific Study of Religion*, vol. 14:4 (1975), pp. 317–30. See also Wayne Proudfoot, *Religious Experience* (Berkeley: University of California Press, 1985).
52. Leuba, *The Psychology of Religious Mysticism.*
53. Reported in W. M. Horton, 'The origin and psychological function of religion, according to Pierre Janet', *American Journal of Psychology*, vol. 35 (1924), p. 20. Leuba's high status as a psychologist of religion continues into the present day. See Benjamin Beit-Hallahmi's estimate in Mircea Eliade (ed.), *Encyclopaedia of Religions*, vol. 8 (New York: Macmillan, 1987), p. 520.
54. Thiselton, *New Horizons in Hermeneutics*, p. 344.

Chapter 8: Modern Scientists Widen the Argument

1. Published in 1851 as *Lectures on the Essence of Religion* (English translation by Ralph Manheim, New York/London: Harper and Row, 1967).
2. *Ibid.*, pp. 219–21. Kant makes a similar assertion in *Religion Within the Limits of Reason Alone* (1793). See the 1960 edition, translated by

Theodore M. Greene and Hoyt H. Hudson (New York: Harper and Row, 1960), p. 163.

3. During the 1960s and 1970s it became fashionable to publish descriptions of one's experience with lysergic acid (LSD) and other hallucinogenic drugs. Predating the 1960s and amongst the most enduringly interesting accounts is that by Aldous Huxley. His essays *The Doors of Perception* and *Heaven and Hell* were published in 1954, well ahead of the mainstream of interest, and reflect on his experiences with mescaline, a substance extracted from the Peyote cactus and traditionally used in religious ceremonies by the Huichol people of Central Mexico.
4. Walter Pahnke, *Drugs and Mysticism: An analysis of the relationship between psychedelic drugs and the mystical consciousness*, PhD dissertation, Harvard University, 1963.
5. See Rick Doblin, 'Pahnke's "Good Friday Experiment": A Long-Term Follow-Up and Methodological Critique', *The Journal of Transpersonal Psychology* 23:1 (1991), pp. 1–28. Reproduced on the Internet at: <http://druglibrary.org/schaffer/lsd/doblin.htm>.
6. *Ibid.*, p. 9 (internet version).
7. *Ibid.*, p. 16.
8. However, the distinction from other spiritual practices may be less than it seems, especially when compared with formal religious exercises. The difference between intentionally taking a chemical substance and intentionally using meditation, with the same religious purpose in each case, is at least debatable. It is conceivable that in either situation an identical physiological response is produced in the body. From what we know of the descriptions people give of their experiences, both involve a heightening of relational consciousness.
9. See B. M. D'Onofrio, L. J. Eaves, L. Murrelle, H. H. Maes, and B. Spilka,'Understanding biological and social influences on religious affiliation, attitudes, and behaviors: a behavior genetic perspective', *Journal of Personality* 67:6 (1999), pp. 953–83.
10. Thomas J. Bouchard Jr, Matt McGue, David Lykken and Auke Tellegen, 'Intrinsic and extrinsic religiousness: genetic and environmental influences and personality correlates', *Twin Research* 2:2 (1999), pp. 88–98.
11. Interestingly, Eaves is not only a professor of genetics; he is also an ordained Anglican priest. He was also one of the collaborators with D'Onofrio in the 'Virginia 30,000' paper.
12. 'Self-transcendence as a measure of spirituality in a sample of older Australian twins', *Twin Research* 2:2 (1999), pp. 81–7. The twins were all aged 50 and over, because this was part of a larger study looking at health in older people.
13. See Dean Hamer, *The God Gene: How Faith is Hardwired into Our Genes* (New York: Doubleday, 2004), p. 49.
14. *Ibid.*
15. See 'Faith boosting genes: a search for the genetic basis of spirituality', *Scientific American*, 27 September 2004.
16. *Ibid.*, p. 56. The ease with which DNA can be collected has been utilised by British bus drivers to dissuade violent attacks from late night drunken

passengers. Apparently hooligans often spit at the drivers, who are equipped with kits to store the saliva and have it analysed as a means of proving the identity of their attacker.

17. Standing for 'Vesicular Monoamine Transporter 2'.
18. There are four types of bases arranged in paired sequences in the core of DNA. Hamer was interested in two of them, Adenine (A) and Cytosine (C).
19. Hamer, *The God Gene,* p. 73.
20. For example, Rhawn Joseph's sprawling edited volume *Neurotheology: Brain, Science, Spirituality, Religious Experience* (San Jose: University Press, 2003).
21. See V. S. Ramachandran and Sandra Blakeslee, *Phantoms in the Brain* (London: Fourth Estate, 1999), ch. 9, 'God and the limbic system', pp. 174–98.
22. What seems to matter here is intensity, for temporal lobe epileptics do not report such experience any more frequently, and possibly less, than the general population. In 1984 Tom Sensky and his colleagues at London University surveyed 137 patients attending neurological clinics in three London teaching hospitals and subdivided them according to whether they were suffering from migraine, primary generalised epilepsy or temporal lobe epilepsy. All were questioned about religious belief and mystical experience. No significant differences were found between the groups. Furthermore, compared to the general population in the United Kingdom they were less likely to report such experience. See Tom Sensky *et al.*, 'The interictal personality traits of temporal lobe epileptics: religious belief and its association with reported mystical experiences', in J. R. Porter *et al.* (eds), *Advances in Epileptology: XVth Epilepsy International Symposium* (New York: Raven Press, 1987), pp. 545–9.
23. See Michael Persinger, 'Religious and mystical experiences as artefacts of temporal lobe function: a general hypothesis', *Perceptual and Motor Skills* 57 (1983), pp. 1235–62.
24. *Ibid.*, p. 1257.
25. There has been one independent attempt to replicate Persinger's experiments. Granqvist and his colleagues at the University of Uppsala concluded that the helmet had no perceptible effect and interpreted Persinger's results as due to suggestibility. See P. Granqvist, M. Fredrikson, P. Unge, A. Hagenfeldt, S. Valind, D. Larhammar, and M. Larsson, 'Sensed presence and mystical experiences are predicted by suggestibility, not by the application of transcranial weak complex magnetic fields', *Neuroscience Letters* 379:1 (2005), pp. 1–6.
26. For an entertaining popular account of a visit to Persinger's laboratory, see John Horgan, *Rational Mysticism: Spirituality meets science in the search for enlightenment* (Boston and New York: Houghton Mifflin Company, 2004), ch. 5, 'The God Machine'.
27. See Eugene d'Aquili and Andrew Newberg's book *The Mystical Mind: Probing the Biology of Religious Experience* (Minneapolis: Fortress Press, 1999).
28. I am most grateful to Dr Beauregard for allowing me to inspect a Powerpoint slide presentation he used in a talk about his work given at a conference in June 2005 in Philadelphia.

29. Now at the University of Wales in Bangor.
30. For a full account of Jackson's fascinating study, see his 1991 DPhil thesis entitled *A Study of the relationship between spiritual and psychotic experience*, available in the library of Oxford University. See also M. C. Jackson and K. W. M. Fulford, 'Spiritual experience and psychopathology', *Philosophy, Psychiatry and Psychology* 4 (1997), pp. 41–90.
31. My colleague Pawel M. Socha of the Institute for the Study of Religion in the University of Krakow comes to similar conclusions in his studies of coping behaviour in times of existential crisis. See D. Hay and P. M. Socha, 'Spirituality as a natural phenomenon: bringing biological and psychological perspectives together', *Zygon* 49:3 (2005), pp. 589–612.
32. But even in purportedly atheistic religions like Theravada Buddhism, local folk religion continues to recognise the reality of 'something there' as an immanent benign presence.
33. Leon Bloy, the French social reformer, liked to say that joy is the most infallible sign of the presence of God.
34. H-G. Gadamer, *Truth and Method*, tr. Joel Weinsheimer and Donald G. Marshall, 2nd revised edn (London: Sheed and Ward, 1989).
35. James Ashbrook, *Where God Lives in the Human Brain* (Naperville, Illinois: Sourcebooks Inc., 2001).
36. See William A. Rottschaefer, 'The image of God of neurotheology: reflections of culturally based religious commitments or evolutionarily based neuroscientific theories?' *Zygon* 34:1 (1999), pp. 57–65.
37. Reported in the *Daily Telegraph*, 20 March 2003.
38. For an exhilarating piece in which Dawkins doesn't waste his energy on scientific detachment, see 'Viruses of the mind', *Free Inquiry*, Summer 1993, pp. 33–41.
39. At the time of writing there are approximately 5,000,000 references to religion as a virus, reported on the internet search engine Google. Others have climbed aboard, e.g. Aaron Lynch, *Thought Contagion: How Ideas Act Like Viruses* (New York: Basic Books, 1998). Possibly falling within this group is Michael Persinger. His public stance is avowedly neutral on the truth or falsity of religious belief, but a flavour of a rather more biased personal agenda can be gathered from the titles of some of his best-known publications, e.g. 'Sense of a presence and suicidal ideation following traumatic brain injury: indications of right-hemispheric intrusions from neuropsychological profiles', *Psychological Reports* 75:3 (1994), Pt I, pp. 1059–70. Also his paper, 'I would kill in God's name: role of sex, weekly church attendance, report of a religious experience and limbic lability', *Perceptual and Motor Skills* 85:1 (1997), pp. 128–130, which, in the era of the suicide bomber might prove to be of importance.
40. See Pascal Boyer, *Religion Explained: The Evolutionary Origins of Religious Thought* (New York: Basic Books, 2001).
41. See Scott Atran, *In Gods We Trust: the Evolutionary Landscape of Religion* (New York: Oxford University Press, 2002).
42. Again, phrases like 'minimally violating ordinary notions of how the world is' take no account of cultural differences of opinion about what ordinary notions are. In some other cultures, Atran's notion of the world would seem abnormal.

43. See Scott Atran and Ara Norenzayan, 'Religion's evolutionary landscape: counterintuition, commitment, compassion, communion', *Behavioral and Brain Sciences* 27: 6 (2004).
44. Wayne Proudfoot, *Religious Experience* (Berkeley: University of California Press, 1987).
45. See Nina P. Azari and Dieter Birnbacher, 'The role of cognition and feeling in religious experience', *Zygon* 39:4 (2004), pp. 901–17.
46. Azari and Birnbacher refer to three errors, the first of which is described as follows: 'Contrary to what the subject thinks, the object is not the full cause of the emotion but only a triggering object or event, the main cause lying in other factors, internal or external to the subject.' Psychoanalytic interpretations of religious experience are an example. In the second type of error: 'Contrary to what the subject thinks, the object does not causally contribute to the emotion but is only a screen on which the emotion is projected. In this case all causal factors lie outside the object, either within or without the subject.' *Ibid.*, p. 914.
47. For a comprehensive review of the modern understanding of the complexity of emotion, see Klaus R. Scherer, Angela Schorr and Tom Johnstone (eds), *Appraisal Processes in Emotion: theory, methods, research* (Oxford: Oxford University Press, 2001).
48. *Ibid.*, p. 912.
49. The atheism of Advaita in India, or in Theravada Buddhism, might be cited as evidence to contradict my thesis, but these forms of atheism are in fact intra-religious. They are aspects of a debate about the nature of transcendence and as such are akin to certain mystical movements in Christianity, for example the near monism of someone like the fourteenth-century Dominican mystic, Meister Eckhart.
50. See John Bowker, *The Sacred Neuron: extraordinary new discoveries linking science and religion* (London/New York: I B Tauris, 2005).
51. See Antonio Damasio, *Descartes' Error: Emotion, Reason and the Human Brain* (New York: Putnam, 1994); *The Feeling of What Happens: Body, Emotion and the Making of Consciousness* (London: Vintage, 2000). Also of interest is Damasio's *Looking for Spinoza: Joy, Sorrow and the Human Brain* (London: William Heinemann, 2003).
52. This point was made very clearly in Jerry Fodor's book, *The Modularity of Mind* (Cambridge, Massachusetts: MIT Press, 1983). Also relevant is the work of the National Professional Development Program in Western Australia. The NPDP made a study to determine whether, in spite of the generally held view of the relativity of values across different cultures, there was an agreed minimum values framework, crossing all cultures in Australia. They did indeed find a large amount of consensus about *ultimate concerns*, *democracy* and *education*. See *Agreed Minimum Values Framework* (Osborne Park: NPDP, Western Australia, 1995).
53. Bowker, *The Sacred Neuron*, p. 118.

Chapter 9: Why Spirituality Is Difficult for Westerners

1. *Wall Street*, starring Michael Douglas in the role of Gordon Gecko. Directed by Oliver Stone, written by Stanley Weiser and Oliver Stone (1987).

2. I saw this slogan advertising Lynx deodorant on the day that the famine in Niger became public knowledge. The conjunction sparked off the following quatrain:

   'Greed is good', the advert said
   In letters three feet high.
   A gaunt child lying almost dead
   In Niger, whispered, 'It's a lie!'

3. Charles Taylor, *Sources of the Self: the Making of the Modern Identity* (Cambridge: Cambridge University Press, 1992).
4. See note 49 in previous chapter.
5. There are many texts on this theme. Possibly the best introduction, because he gives a systematic overview of its many dimensions, is Stephen Lukes' *Individualism*, published in the series *Key Concepts in the Social Sciences* (Oxford: Blackwell, 1973); See also Colin Morris, *The Discovery of the Individual, 1050–1200* (London: SPCK, 1972); Louis Dumont, *Essays on Individualism* (Chicago: University of Chicago University Press, 1986); Aaron Gurevich, who disagrees with Morris' claim that individualism appeared in the twelfth century; see his *The Origins of European Individualism* (Oxford: Blackwell, 1995).
6. For further information on these questions see the articles by Terrence Deacon on 'Biological aspects of language' (pp. 128–33) and C. B. Stringer on 'Evolution of early humans' (pp. 241–51) in Steve Jones, Robert Martin and David Pilbeam (eds), *The Cambridge Encyclopedia of Human Evolution* (Cambridge: Cambridge University Press, 1994).
7. The question of the self-awareness of animals is hotly disputed. It is discussed at length in Marc Bekoff *et al.* (eds), *The Cognitive Animal* (Cambridge, Massachusetts: MIT Press, 2002).
8. See ch. 6.
9. For a popular discussion of the effect of language on self-awareness, see John McCrone, *The Ape that Spoke: Language and the Evolution of the Human Mind* (London: Picador, 1990).
10. See Alexander Luria's book *Cognitive Development: Its Cultural and Social Foundations,* tr. Martin Lopez-Morillas and Lynn Solotaroff, edited by Michael Cole (Cambridge, Massachusetts: Harvard University Press, 1976). Because of difficulties with Stalinist censorship these findings were not published in the Soviet Union until the 1970s.
11. One only has to think of the way that reading and writing dominate our everyday lives, now added to by the ubiquity of the Internet and the world wide web, to begin to see that the mode of action of our consciousness is very different from that of our non-literate forebears. See for example, John L. Locke's book, *Why We Don't Talk to Each Other Any More: the Devoicing of Society* (New York: Simon and Schuster, 1998).
12. Note for instance the experience of the psychotherapist Eugene Gendlin (1981), when encountering academically high-flying clients in his Chicago consulting rooms. Gendlin comments on the disconcerting fact that he was unable to help many of them to explore their immediate emotional difficulties because they were isolated from the felt sense of their bodies. Too good a training in academic detachment had crippled them. See also the related arguments from the eminent neurologist Antonio

Damasio in *Descartes' Error: Emotion, Reason and the Human Brain* and *The Feeling of What Happens: Body, Emotion and the Making of Consciousness* on the importance of the body in relation to emotion and consciousness.

13. Available in Penguin Classics, London, 1990.
14. See Max Weber, *The Protestant Ethic and the Spirit of Capitalism*, tr. Talcott Parsons (London: George Allen and Unwin, 1930), p. 104. Pastoral need led to the mitigation of the doctrine and it became accepted that one plausible sign of election was material prosperity in this life. Weber's (often disputed) contention was that this belief encouraged the growth of capitalism in Europe.
15. The teaching of Cornelius Jansen, which split the Roman Catholic Church in France in the mid-seventeenth century. Jansen emphasised the belief that an individual can do nothing to assure their own salvation: all is due to divine grace. Jansenism was centred on the abbey of Port Royal and Pascal was its most prominent lay supporter. The Jansenists were excommunicated in 1719.
16. Reprinted in H. C. Clark, *Commerce, Culture and Liberty: Readings on Capitalisation Before Adam Smith* (Indianapolis: Liberty Fund, 2003).
17. See John Macmurray, *The Self as Agent*, with an introduction by Stanley M. Harrison (London: Faber and Faber, 1995), p. 71.
18. See David Berman's fascinating thesis on hidden atheism in *A History of Atheism in Britain: From Hobbes to Russell* (London and New York: Routledge, 1990).
19. For a readable account of orthodox altruism theory, see Helena Cronin, *The Ant and the Peacock* (Cambridge: Cambridge University Press, 1993).
20. In Jean Hampton, *Hobbes and the Social Contract Tradition* (Cambridge: Cambridge University Press, 1988)
21. In *Philosophical Rudiments concerning Government and Society*, ch. 1, Section 4, pp. 25–26, quoted in C. B. Macpherson, *The Political Theory of Possessive Individualism* (Oxford: Oxford University Press, 1962), p. 44.
22. Thomas Hobbes, *Leviathan*, edited with an introduction by C. B. Macpherson (London: Penguin Classics, 1985).
23. *Ibid.*, p. 188.
24. *Ibid.*, p. 228.
25. Quoted in Hampton, *Hobbes and the Social Contract Tradition*, p. 10.
26. C. B. Macpherson, *The Political Theory of Possessive Individualism*. op. cit.
27. Albert Hirschman, *The Passions and the Interests: Political Arguments for Capitalism before its Triumph* (first published Princeton: Princeton University Press, 1977; republished as a twentieth anniversary edition with a foreword by Amartya Sen, 1997).
28. Hirschman, *The Passions and the Interests*, p. 14.
29. *Ibid.*, p. 40.
30. Quoted in Hirschman, *Ibid.*, p. 50.
31. Adam Smith, *The Wealth of* Nations, currently available in the two-volume Penguin edition, with an introduction and notes by Andrew Skinner (London: Penguin, 1999).
32. A distinction must be made between Smith's account of the way things are in capitalist society and his personal view of ethics. Smith's moral

philosophy is expounded in *The Theory of Moral Sentiments* (1759) published 17 years before *The Wealth of Nations.* He has much to say of 'sympathy' which suggests that it is not remote from relational consciousness. The apparent ethical disjunction between the two works has led to much discussion. It must be added that Smith's rhetoric, particularly in the later chapters of *The Wealth of Nations* frequently makes clear his distaste for some of the situations he is describing (See J. Z. Muller, *Adam Smith in His Time and Ours: Designing the Decent Society* (New York: The Free Press (Macmillan), 1993).

33. English translation by Ralph Manheim, published by Harper & Row in New York in 1967.
34. Karl Marx, *Theses on Feuerbach,* first published in 1845. Reprinted in *Marx and Engels on Religion* (Moscow: Progress Publishers, 1957).
35. See Van A. Harvey, *Feuerbach and the Interpretation of Religion* (Cambridge: Cambridge University Press, 1977).
36. Pseudonym of Johann Caspar Schmidt.
37. Max Stirner, *The Ego and Its Own,* tr. Steven Byington, with an introduction by Sydney Parker (London: Rebel Press, 1993).
38. *Ibid.*, p. 5.
39. *Ibid.*, pp. 296–7. His lover in Berlin left him in disgust, accusing him appropriately enough of being totally self-centred. She eventually entered religious life and died in a convent.
40. R. W. K. Paterson, *The Nihilistic Egoist Max Stirner,* published for the University of Hull (Oxford: Oxford University Press, 1971).
41. Bruno Bauer, another member of the Young Hegelian group in Berlin and a former theologian.
42. Paterson, *The Nihilistic Egoist Max Stirner,* p. 31.
43. *Ibid.*, p. 263.
44. *Ibid.*, p. 197.
45. Quoted in Michael Walzer 'The communitarian critique of liberalism', *Political Theory,* 18:1 (1990), pp. 6–23.
46. Karl Marx, *The German Ideology.* An edition was published in the UK (Prometheus Books, 1998).
47. For example, Francis Wheen in his highly readable *Karl Marx* (London: Fourth Estate, 2000).
48. Hobbes may have dispensed with religion, but it would be interesting to investigate the theological complexion of his early upbringing. He certainly encountered Calvinist opinions when he was a student at Magdalen Hall in Oxford and this may have encouraged in him a belief in the natural depravity of the species. When he discarded religious belief in his maturity he would then have been left with depravity, now deprived of saving grace.
49. See Martin Buber, *Between Man and Man,* tr. Ronald Gregor Smith (London: Fontana, 1961), p. 64.
50. *Ibid.*, p. 66.
51. *Ibid.*, p. 61.
52. See *Bowling Alone: The Collapse and Revival of American Community* (New York: Simon and Schuster, 2000).
53. The phrase comes from the sociologist Philip Selznick. See his book *The*

Moral Commonwealth: Social Theory and the Promise of Community (Berkeley: University of California Press, 1992).

54. For useful and contrasting accounts of the history, see: Berman, *A History of Atheism in Britain*; Michael Buckley, *At the Origins of Modern Atheism* (New Haven and London: Yale University Press, 1987); Samuel Preus, *Explaining Religion: Criticism and Theory from Bodin to Freud* (New Haven and London: Yale University Press, 1987).
55. Buckley, *At the Origins of Modern Atheism.*

Chapter 10: The Problems of the Institution

1. Philip Larkin, from 'Church Going', *Collected Poems,* edited with an introduction by Anthony Thwaite (London: Faber, 1988), p. 97.

Chapter 11: Treating the Sickness of the Spirit

1. John Calder, in *The Philosophy of Samuel Beckett* (London: Calder Publications, 2001), p. 3.
2. Marx's response to the question, 'What is your favourite maxim?' in the Victorian parlour game 'Confessions'.
3. *Big Fish,* directed by Tim Burton (2003).
4. Michael Billington, *Guardian,* 26 August 2005.
5. See James Knowlson's *Damned to Fame: The Life of Samuel Beckett*; and Anthony Cronin's *Samuel Beckett, the Last Modernist* (op. cit. chapter 1).
6. The title of Søren Kierkegaard's exploration of despair: *The Sickness Unto Death: A Christian Psychological Exposition of Edification and Awakening by Anti-Climacus,* tr. Alastair Hannay (London: Penguin Books, 1989).
7. For a discussion of Marxist ethics that takes up the essentialist position, see Norman Geras, *Marx and Human Nature: Refutation of a Legend* (London: Verso, 1983); also Lawrence Wilde, *Ethical Marxism and its Radical Critics* (London: Macmillan Press, 1998). I am grateful to Larry Wilde for his advice on Marx's ethics and for referring me to the above texts.
8. Nancy Bancroft, 'Spirituality in Marxism: a Communist view', ch. 2 in Benjamin B. Page (ed.), *Marxism and Spirituality: An International Anthology* (Westport, Connecticut/London: Bergin and Garvey, 1993).
9. Stirner, *The Ego and Its Own,* ch. 10.
10. N. Lobkowicz, 'Karl Marx and Max Stirner' in Frederick J. Adelmann (ed.), *Demythologizing Marxism* (Beacon Hill: Boston College, 1969). Not all students of Marx agree that he lost his belief in essence. The moral fervour of his battle on behalf of the downtrodden surely suggests they are right. Marx himself had warned against a superstitious trust in historical inevitability, 'History does *nothing* … history is *nothing but* the activity of man pursuing his own aims.' Furthermore he continued to speak of the 'alienation' of the workers, not merely from the products of their labours but also implicitly from their true selves. But the muffling of Marx's belief in human essence meant that concrete political attempts to change the world in accordance with his understanding have all too often been blind to what I have called relational consciousness.
11. Devastatingly criticised by Karl Popper in his essay, *The Poverty of Historicism* (London: Routledge and Kegan Paul, 1957).

12. See Note 2.
13. See David Tacey, *The Spirituality Revolution: the emergence of contemporary spirituality* (Hove and New York: Brunner-Routledge, 2000).
14. Levinas is both very important and, for someone not acquainted with continental philosophy, often maddeningly opaque. See *The Levinas Reader* edited by Sean Hand (Oxford: Blackwell, 2000). I find it easier to approach Levinas at second hand, for example in Zygmunt Bauman's *Postmodern Ethics* (Oxford: Blackwell, 1993). I also owe a debt of gratitude to Michael Barnes SJ of Heythrop College, London University for helpful advice.
15. See Dana Zohar and Ian Marshall, *Spiritual Capital: Wealth We Can Live By* (London: Bloomsbury, 2004); also Laurence R. Iannaccone and Jonathan Klick, 'Spiritual Capital: An Introduction and Literature Review', preliminary draft, prepared for the Spiritual Capital Planning Meeting, 9–10 October 2003, Cambridge, Massachusetts. Available at <http://www.metanexus.net/spiritual_capital/pdf/review.pdf>.
16. The ethical importance of relational consciousness, and the unsatisfactory nature of current zoological explanations of ethical behaviour are discussed in greater detail in *The Spirit of the Child*, especially ch. 8. There I draw attention to the weakness of the 'kin selection' and 'reciprocal altruism' theories that currently dominate the debate. Both of them assume a calculative manipulation of the situation that is alien to true altruism, and which is not required if we take relational consciousness into account. See also in this connection, Kristen Renwick Monroe's fine book *The Heart of Altruism: Perceptions of a Common Humanity* (Princeton: Princeton University Press, 1996).
17. Including that new–old arrival, Islam.
18. Tacey, *The Spirituality Revolution*, p. 31.
19. See Karl Rahner SJ, 'The experience of God today' in *Theological Investigations I*, tr. David Bourke (London: Darton, Longman and Todd, 1974), pp. 149–65.
20. See Carl R. Rogers, *Client Centred Therapy: Its Current Practice, Implications and Theory* (London: Constable and Robinson, 2003).
21. The clearest description of the here-and-now practice developed by Wilfred Bion is in his book, *Experiences in Groups*.
22. See for example, Gene Gendlin's, *Experiencing and the Creation of Meaning* (Chicago: Northwestern University Press, 1997). Also his popular paperback, *Focusing: how to open up your deeper feelings and intuition* (London: Rider, 25th anniversary edition, 2003).
23. See Peter Campbell and Edwin McMahon, *Biospirituality: Focusing as a Way to Grow* (Chicago: Loyola University Press, 1997). Also consult the Biospirituality website at <http://www.biospiritual.org/index.html>.
24. See Alfred Schutz, 'Making music together: a study in social relationship' in Arvid Brodersen (ed.), *Collected Papers II: Studies in Social Theory* (The Hague: Martinus Nijhoff, 1964), pp. 135–58.
25. See Mihaly Csikszentmihalyi, *Beyond Boredom and Anxiety* (San Francisco: Jossey-Bass, 1975); also Mihaly Csikszentmihalyi, Kevin Rathunde and Samuel Whalen, *Talented Teenagers: The Roots of Success and Failure* (Cambridge: Cambridge University Press, 1993).

26. See Mary Jo Neitz and James V. Spickard, 'Steps toward a sociology of religious experience: the theories of Mihaly Csikszentmihalyi and Alfred Schutz', *Sociological Analysis* 51:1 (1990), pp. 15–33.
27. See Isabella Csikszentmihalyi, 'Flow in a historical context: the case of the Jesuits' in Mihaly Csikszentmihalyi and Isabella Csikszentmihalyi (eds), *Psychological Studies of Flow in Consciousness* (New York: Cambridge University Press, 1988), pp. 232–48.
28. See Roberto Assagioli's *Psychosynthesis: a Manual of Techniques* (London: HarperCollins, 1999); also Piero Ferrucci's rather more readable, *What We May Be: Techniques for Psychological and Spiritual Growth* (Los Angeles: J. P. Tarcher, 1982); and Jean Hardy, *A Psychology with a Soul* (London: Woodgrange Press, 1996).
29. See Frederick S. Perls, *Gestalt Therapy: Excitement and Growth in the Human Personality* (London: Souvenir Press, 1994).
30. For an interesting account, see Roger Shattuck's *Forbidden Knowledge: from Prometheus to Pornography* (San Diego/New York/London: Harvest Books, 1997).
31. 1 Corinthians 8:1.
32. See Sabrina Petra Ramet (ed.), *Religious Policy in the Soviet Union* (Cambridge: Cambridge University Press, 1992).
33. George Lakoff and Mark Johnson's *Metaphors We Live By* (Chicago: University of Chicago Press, 1980) is still an eye-opening and amusing account of the pervasiveness of metaphor.
34. *Ibid.*, pp. 7–8.
35. See Paul Ricœur, *The rule of metaphor: multi-disciplinary studies of the creation of meaning in language*, tr. Robert Czerny (London: Routledge and Kegan Paul, 1978); Sally McFague, *Metaphorical Theology: Models of God in Religious Language* (London: SCM Press, 1982); Janet Martin Soskice, *Metaphor and Religious Language* (Oxford: Clarendon Press, 1987).
36. See Thomas Kuhn's pioneering book, *The Structure of Scientific Revolutions* (Chicago: University of Chicago Press, 1962). The heated debate aroused by Kuhn is reflected in Imre Lakatos and Alan Musgrave (eds), *Criticism and the Growth of Knowledge* (Cambridge: Cambridge University Press, 1970). See also Ian Barbour *Myths, Models and Paradigms* (New York: Harper and Row, 1974).
37. Edward Evans-Pritchard, in *Theories of Primitive Religion* (Oxford University Press, 1965), p. 13.
38. On the day that I wrote this sentence, in an interview in the *Guardian* newspaper the author Philip Roth responded to a question about his attitude to religion, saying, 'I don't even want to talk about it, it's not interesting to talk about the sheep referred to as believers', *Guardian G2*, 14 December 2005, p. 17.
39. Critically reviewed by the Finnish psychologist of religion Kalevi Tamminen in 'Religious experiences in childhood and adolescence: a viewpoint of religious development between the ages of 7 and 20', *International Journal for the Psychology of Religion* 4:2 (1994), pp. 61–85.
40. For a discussion of the appropriateness of metaphors of God's immanence from the perspective of science, see Arthur Peacocke's *Paths from*

Science towards God (Oxford: Oneworld Publications, 2001).

41. See Benjamin Beit-Hallahmi and Michael Argyle, *The Psychology of Religious Behaviour, Belief and Experience* (London and New York: Routledge, 1997), pp. 27–31, 244–7.
42. Trotter lectured in theology at St Andrews University for many years, following his time in Peckham. As an illustration of what he means, Trotter ambitiously chooses to explore the correspondence he sees between the science of human ethology and the theological investigation of the Christian doctrine of the Trinity. My undergraduate education was in zoology and out of that background I have a memory of trying in vain to make sense of a seminary text on the Trinity, lent to me by a priest friend. I found it to be like eating dust, so Trotter's choice of theme to illustrate his correspondence idea struck me as ambitious to the point of being foolhardy. Nevertheless I think he succeeds in making his point. His argument centres on the claim that trinitarian doctrine grew out of the eucharistic life of the early church. That is to say the doctrine is not in the least like a theory accepted at second hand from some external ecclesiastical authority. It is literally a response to the practical human experience of communion, and as such is an insight into the nature of human life at its richest, thus connecting his argument with the science of *ethology* or animal behaviour. Trotter gives the term a richer meaning by reminding us that the Greek root is *ethos*, referring to the full life, the life one aspires to.
43. See I. H. Pearse and L. H. Crocker, *The Peckham Experiment: a study in the living structure of society* (London: George Allen and Unwin Ltd, 1943).
44. Douglas Trotter, *Wholeness and Holiness: A Study in Human Ethology and the Holy Trinity*, (Glasgow: Pioneer Health Foundation, 2003).
45. *Ibid.*, p. 188.
46. See Muneo Jay Yoshikawa, 'The double swing model of intercultural communication between the East and the West', in D. Lawrence Kincaid (ed.), *Communication Theory: Eastern and Western Perspectives* (San Diego: Academic Press, 1987).
47. Well-known atheist missionaries like Richard Dawkins face analogous temptations.
48. Reported in Richard Rubenstein's powerful book *After Auschwitz: History, Theology and Contemporary Judaism*, 2nd edn (Baltimore and London: The Johns Hopkins University Press, 1992).
49. *Ibid.*, p. 3.
50. George Steiner, *The Death of Tragedy* (London: Faber and Faber, 1961).
51. See George Steiner, ' "Tragedy" reconsidered', *New Literary History* 35 (2004), 1–15, p. 15.
52. For a fascinating reinterpretation of divine immanence that illuminates this perspective, see Arthur Peacocke, *Paths from Science Towards God.*
53. For a stimulating review of the complex moral and political issues raised by globalisation, see Joseph Stiglitz's *Globalization and its Discontents* (London: Penguin Books, 2002).
54. Honorary Professor in the Centre for Organizational Renewal at Cranfield University. I am grateful to Dr Lawrence for permitting me to examine an unpublished chapter of a forthcoming book on this subject. Many of the

insights in the final part of the present chapter are due to him.

55. See John D. Barrow and Frank J. Tipler, *The Anthropic Cosmological Principle* (Oxford: Oxford Paperbacks, 1988).
56. In *The Unconscious Civilisation* (London: Penguin, 1998), Ralston Saul argues for a clear division between the economic and political, maintaining that economic structures such as capitalism possess of themselves no mechanisms for the protection of people's rights, the environment, or the public interest. This of course follows from the collapse of ethics into economic prudence.

An Unscientific Postscript

1. I am grateful to Rabbi Soetendorp for giving me permission to reproduce the substance of these two narratives.
2. This is a complex and unresolved issue. For differing points of view, see Christopher Browning, *Ordinary Men: Reserve Police Battalion 101 and the Final Solution in Poland* (London: Penguin Books, 2001); and Daniel Goldhagen, *Hitler's Willing Executioners: Ordinary Germans and the Holocaust* (London: Abacus, 1997).
3. Primo Levi, *If This Is a Man* (London: Abacus Press, 1991).
4. United States Holocaust Memorial Museum's website: <http://www.ushmm.org/wlc/article.php?lang=enand ModuleId=10005537>
5. Browning, *Ordinary Men.*
6. A major feature of this story is the fact that the person who acted as a neighbour to the wounded man was a Samaritan, a member of a despised minority.
7. Luke 10:29–37.

Bibliography

Acquaviva, S., *Eros morte ed esperienza religiosa* (*Eros, Death and Religious Experience*) (Bari: Laterza, 1991).

Adorno, T., Frenkel-Brunswik, E., Levinson, D. and Nevitt Sanford, R., *The Authoritarian Personality* (New York: Harper and Row, 1950).

Ahern, G., *Spiritual/Religious Experience in Modern Society* (Oxford: RERU, 1990).

Albright, C. and Ashbrook, J., *Where God Lives in the Human Brain* (Naperville, Illinois: Sourcebooks Inc., 2001).

Allport, G., *The Individual and His Religion* (New York: Macmillan, 1962).

American Psychiatric Association, *Diagnostic and Statistical Manual of Mental Disorders* (DSM-3, 3rd edn, revised) (Washington DC: APA, 1987).

Anon., *The Cloud of Unknowing*, tr. Clifton Wolters (London: Penguin Books, 1961).

Arnold, M., *Poems*, with an introduction by Kenneth Allott (London: Dent, 1965).

Arvon, H., *Aux Sources de l'Existentialisme: Max Stirner* (Paris: PUF, 1954).

Assagioli, R., *Psychosynthesis: a Manual of Techniques* (London: HarperCollins, 1999).

Atran, S., *In Gods We Trust: the Evolutionary Landscape of Religion* (New York: Oxford University Press, 2002).

—and Norenzayan, A., 'Religion's evolutionary landscape: counterintuition, commitment, compassion, communion', *Behavioral and Brain Sciences* 27:6 (2004), pp. 713–30.

Augustine, St, *Confessions*, tr. Henry Chadwick (London: Penguin, 1991).

Aunger, R. (ed.), *Darwinising Culture: the Status of Memetics as a Science* (Oxford: Oxford University Press, 2001).

Azari, N. P. and Birnbacher, D., 'The role of cognition and feeling in religious experience', *Zygon* 39:4 (2004), pp. 901–17.

Bancroft, N., 'Spirituality in Marxism: a Communist view', in Benjamin B. Page (ed.), *Marxism and Spirituality: An International Anthology* (Westport, Connecticut/London: Bergin and Garvey, 1993).

Barbour, I., *Myths, Models and Paradigms* (New York: Harper and Row, 1974).

Barrow, J. D. and Tipler, F. J., *The Anthropic Cosmological Principle* (Oxford: Oxford Paperbacks, 1988).

Bauman, Z., *Postmodern Ethics* (Oxford: Blackwell, 1993).

Beardsworth, T., *A Sense of Presence* (Oxford: RERU, 1977).

Beckett, S., *The Unnameable* (London: John Calder, 1959).

—*Waiting for Godot* (London: Faber and Faber, 1956).

Beit-Hallahmi, B., article on James H. Leuba, in Mircea Eliade (ed.),

Encyclopaedia of Religions, vol. 8 (New York: Macmillan, 1987).
—and Argyle, M., *The Psychology of Religious Behaviour, Belief and Experience* (London and New York: Routledge, 1997).
Bekoff, M., Allen, C. and Burghardt, G. M. (eds), *The Cognitive Animal: Empirical and Theoretical Perspectives on Animal Cognition* (Cambridge, Massachusetts: MIT Press, 2002).
Berman, D., *A History of Atheism in Britain: from Hobbes to Russell* (London and New York: Routledge, 1990).
Billington, M., review of *Waiting for Godot*, *Guardian*, 26 August 2005.
Bion, W., *Experiences in Groups* (London: Tavistock Press, 1961).
Blackmore, S., *The Meme Machine* (Oxford: Oxford University Press, 1999).
Bouchard, T., McGue, M., Lykken, D. and Tellegen, A., 'Intrinsic and extrinsic religiousness: genetic and environmental influences and personality correlates', *Twin Research* 2:2 (1999), pp. 88–98.
Bowker, J., *The Sacred Neuron: Extraordinary New Discoveries Linking Science and Religion* (London/New York: I. B. Taurus, 2005).
Boyer, P., *Religion Explained: The Evolutionary Origins of Religious Thought* (New York: Basic Books, 2001).
Bradburn, N., *The Structure of Psychological Wellbeing* (Chicago: Aldine Press, 1969).
Brierley, P., *Religious Trends No. 1: 1999/2000* (London: Christian Research Association, 2000).
—*The Tide is Running Out* (London: Christian Research Association, 2000).
Brown, C., *The Death of Christian Britain: Understanding Secularisation 1800–2000* (London and New York: Routledge, 2001).
Browning, C., *Ordinary Men: Reserve Police Battalion 101 and the Final Solution in Poland* (London: Penguin Books, 2001).
Bruce, S., *God Is Dead: Secularization in the West* (Oxford: Blackwell, 2002).
Bryden, M., *Samuel Beckett and the Idea of God* (London: Macmillan, 1998).
Buber, M., *Between Man and Man*, tr. Ronald Gregor Smith (London: Fontana, 1961).
Buckley, M., *At the Origins of Modern Atheism* (New Haven and London: Yale University Press, 1987).
Burckhardt, J., *The Civilisation of the Renaissance in Italy* (London: Penguin Classics, 1990).
Calder, J., *The Philosophy of Samuel Beckett* (London: Calder Publications, 2001).
Campbell, P. and McMahon, E., *Biospirituality: Focusing as a Way to Grow* (Chicago: Loyola University Press, 1997).
Carpenter, W. B., *Mental Physiology* (New York: D. Appleton and Co., 1881).
Caussade, J-P. de, *Self Abandonment to Divine Providence*, tr. Algar Thorold (London: Burns and Oates, 1933).
Chapman, J., *Spiritual Letters* (London and Sydney: Sheed and Ward, 1935).
Clark, H. C. (ed), *Commerce, Culture and Liberty: Redings on Capitalism Before Adam Smith* (Indianapolis: Liberty Fund, 2003).
Clough, A., *The Poems of Arthur Hugh Clough*, 2nd edn, ed. F. L. Mulhauser (Oxford: Clarendon Press, 1974).
Coe, G., *The Spiritual Life* (New York: Fleming H. Revell Co., 1900).
—*The Psychology of Religion* (Chicago: University of Chicago Press, 1916).

Cohen, J. M. and Phipps, J. F., *The Common Experience* (London: Rider, 1979).
Comte, A., *The Positive Philosophy of Auguste Comte*, vol. I, tr. Harriet Martineau (London: Batoche Books, 2000/1853).
Cousineau, T., *Waiting for Godot: Form in Movement* (Boston: Twayne Publishers, 1990).
Crick, F., interview reported in *Daily Telegraph*, 20 March 2003.
Cronin, A., *Samuel Beckett: the Last Modernist* (London: HarperCollins, 1996).
Cronin, H., *The Ant and the Peacock* (Cambridge: Cambridge University Press, 1993).
Csikszentmihalyi, I., 'Flow in a historical context: the case of the Jesuits' in I. Csikszentmihalyi and M. Csikszentmihalyi (eds), *Psychological Studies of Flow in Consciousness* (New York: Cambridge University Press, 1988/1999).
Csikszentmihalyi, M., *Beyond Boredom and Anxiety* (San Francisco: Jossey-Bass, 1975).
—Rathunde, K. and Whalen, S., *Talented Teenagers: The Roots of Success and Failure* (Cambridge: Cambridge University Press, 1993).
d'Aquili, E. and Newberg, A. B., *The Mystical Mind: Probing the Biology of Religious Experience* (Minneapolis: Fortress Press, 1999).
d'Onofrio, B. M., Eaves, L. J., Murrelle, L., Maes, H. H. and Spilka, B., 'Understanding biological and social influences on religious affiliation, attitudes, and behaviors: a behavior genetic perspective', *Journal of Personality* 67:6 (1999), pp. 953–83.
Damasio, A., *Descartes' Error: Emotion, Reason and the Human Brain* (New York: Putnam, 1994).
—*The Feeling of What Happens: Body, Emotion and the Making of Consciousness* (London: Vintage, 2000).
—*Looking for Spinoza: Joy, Sorrow and the Human Brain* (London: William Heinemann, 2003).
Darlington, C. D., *The Evolution of Man and Society* (London: Allen and Unwin, 1969).
Davie, G., *Religion in Britain since 1945: Believing without Belonging*, with an introduction by David Martin (Oxford: Blackwell, 1994).
Dawkins, R., *The Selfish Gene* (Oxford University Press, 1976).
—*The Blind Watchmaker* (London: Longman Scientific and Technical, 1986).
—'Viruses of the mind', *Free Inquiry*, Summer 1993, pp. 33–41.
Deacon, T., 'Biological aspects of language' in *The Cambridge Encyclopedia of Human Evolution*, Steve Jones, Robert Martin and David Pilbeam (eds) (Cambridge: Cambridge University Press, 1994), pp. 128–33.
Doblin, R., 'Pahnke's "Good Friday Experiment": a long-term follow-up and methodological critique', *The Journal of Transpersonal Psychology* 23:1 (1991), pp. 1–28. Reproduced on the Internet at: <http://druglibrary.org/schaffer/lsd/doblin.htm>
Donajgrodski, A. (ed.), *Social Control in Nineteenth-century Britain* (London: Croom Helm, 1977).
Donaldson, M., *Human Minds* (London: Allen Lane, 1992).
Donne, J., *The Epithalamions, Anniversaries and Episodes*, ed. W. Milgate (Oxford: Clarendon Press, 1978).
Dreyfus, H., *Being-in-the-World* (Cambridge, Massachusetts: MIT Press, 1991).

Duckworth, C. (ed.), *Samuel Beckett: En attendant Godot* (London: Harrap, 1966).
Dumont, L., *Essays on Individualism* (Chicago: University of Chicago Press, 1986).
Durham, W., *Corvolution: Genes, Cultures and Human Destiny* (Stanford University Press, 1991).
Durkheim, É., *The Elementary Forms of the Religious Life*, tr. J. W. Swain (London: Allen and Unwin, 1913).
Eagleton, T., *The Illusions of Postmodernism* (Oxford: Blackwell, 1996).
Edwards, J., *The Religious Affections* (Edinburgh: Banner of Truth Trust, 1986/1746).
Eliade, M. (ed.), *The Encyclopedia of Religion*, vol. 6 (New York: Macmillan, 1986).
Engels, F., 'Letter from Wuppertal', under the pseudonym Friedrich Oswald, reprinted in Marx and Engels, *Collected Works*, vol. II (London: Lawrence and Wishart, 1975).
Evans-Pritchard, E. E., *Theories of Primitive Religion* (Oxford: Oxford University Press, 1965).
Ferrucci, P., *What We May Be: Techniques for Psychological and Spiritual Growth* (Los Angeles: J. P. Tarcher, 1982).
Feuerbach, L., *The Essence of Christianity*, tr. George Eliot with an introduction by Karl Barth (New York: Harper Torchbooks, 1957).
—*Lectures on the Essence of Religion*, tr. Ralph Manheim (New York: Harper and Row, 1967/1851).
Fodor, J., *The Modularity of Mind* (Cambridge, Massachusetts: MIT Press, 1983).
Freud, S., 'A religious experience', reprinted in the *Standard Edition of the Works of Sigmund Freud*, vol. XXI (London: Hogarth Press, 1928/1959).
Gadamer, H-G., *Truth and Method* (London: Sheed and Ward, 1989).
Gallup Poll, USA, '4 in 10 Americans have had unusual spiritual experiences', Press release, 1985.
Gardavsky, V., *God Is Not Yet Dead* (London: Penguin Books, 1973).
Gardner, H. (ed.), *The Faber Book of Religious Verse* (London: Faber and Faber, 1972).
Geertz, C., *The Interpretation of Cultures* (New York: Basic Books Inc., 1973).
—'The way we think now: toward an ethnography of modern thought', in *Local Knowledge: Further Essays in Interpretive Anthropology* (New York: Basic Books, 1983).
Gendlin, E., *Experiencing and the Creation of Meaning* (Chicago: Northwestern University Press, 1997).
—*Focusing: how to open up your deeper feelings and intuition*, 25th anniversary edition (London: Rider, 2003).
Geras, N., *Marx and Human Nature: Refutation of a Legend* (London: Verso, 1983).
Ginger, R., *Eugene Debs: A Biography* (New York: Simon and Schuster, 1962).
Glock, C. and Stark, R., *Religion and Society in Tension* (Chicago: Rand McNally, 1965).
Godin, A., *The Psychological Dynamics of Religious Experience* (Birmingham, Alabama: Religious Education Press, 1985).

Goldhagen, D., *Hitler's Willing Executioners: Ordinary Germans and the Holocaust* (London: Abacus, 1997).
Gopnik, A., *Guardian G2*, Friday 7 January 2005.
—Meltzoff, A. and Kuhl, P., *The Scientist in the Crib: Minds, Brains and How Children Learn* (New York: HarperCollins, 2001).
Gordon, F. M., 'The debate between Feuerbach and Stirner: an introduction', *The Philosophical Forum* 8 (1976), 2-3-4, reproduced at: <http://www.nonserviam.com/egoistarchive/stirner/articles/gordon.html>
Graham, A., *Conversations: Christian and Buddhist* (New York: Harcourt Brace Jovanovich, 1968).
Granqvist, P., Fredrikson, M., Unge, P., Hagenfeldt, A., Valind, S., Larhammar, D., and Larsson, M., 'Sensed presence and mystical experiences are predicted by suggestibility, not by the application of transcranial weak complex magnetic fields', *Neuroscience Letters* 379:1 (2005), pp. 1–6.
Greeley, A., *The Sociology of the Paranormal: A Reconnaissance*, Sage Research Papers in the Social Sciences, Studies in Religion and Ethnicity Series no. 90–023 (Beverley Hills/London: Sage Publications, 1975).
Guardian: 'Spirit', *Guardian Weekend*, 28 February 2004, pp. 41–55.
Gurevich, A., *The Origins of European Individualism* (Oxford: Blackwell, 1995).
Hamer, D., *The God Gene: How Faith is Hardwired into Our Genes* (New York: Doubleday, 2004).
Hampton, J., *Hobbes and the Social Contract Tradition* (Cambridge: Cambridge University Press, 1988).
Hand, S. (ed.), *The Levinas Reader* (Oxford: Blackwell, 1989).
Hardy, A., *The Living Stream: A Restatement of Evolution Theory and its Relation to the Spirit of Man* (London: Collins, 1965).
—*The Divine Flame* (London: Collins, 1966).
—*The Biology of God* (London: Jonathon Cape, 1975).
—*The Spiritual Nature of Man* (Oxford: Clarendon Press, 1979).
Hardy, A., Koestler, A. and Harvie, R., *The Challenge of Chance* (London: Hutchinson, 1973).
Hardy, J., *A Psychology with a Soul* (London: Woodgrange Press, 1996).
Harries, R., *Questioning Belief* (London: SPCK, 1995).
Harvey, Van A., *Feuerbach and the Interpretation of Religion* (Cambridge: Cambridge University Press, 1997).
Hay, D., 'Religious experience amongst a group of postgraduate students: a qualitative study', *Journal for the Scientific Study of Religion* 18 (1979), pp. 164–82.
—'Asking questions about religious experience', *Religion* 18 (1988), pp. 217–29.
—'"The biology of God": what is the current status of Hardy's hypothesis?', *International Journal for the Psychology of Religion* 4:1 (1994), pp. 1–23.
—'Memories of a Calvinist childhood', in W. Gordon Lawrence (ed.), *Roots in a Northern Landscape: Celebrations of Childhood in the North East of Scotland* (Edinburgh: Scottish Cultural Press, 1996).
—'Psychologists interpreting conversion: two American forerunners of the hermeneutics of suspicion', *History of the Human Sciences* 12:1 (1999), pp. 55–72.

—'Why is implicit religion implicit?', *Implicit Religion* 6:1 (2003), pp. 17–41.
—'A biologist of God: Alister Hardy in Aberdeen', *Aberdeen University Review* LX 3, no. 211 (2004), pp. 208–21.
—and Hammond, J., ' "When you pray, go to your private room": a reply to Adrian Thatcher', *British Journal of Religious Education* 14:3 (1992).
—and Heald, G., 'Religion is good for you', *New Society*, 17 April 1987.
—and Hunt, K., *The Spirituality of People Who Don't Go to Church*, final report (Adult Spirituality Project: Nottingham University, 2000).
—and Morisy, A., 'Reports of ecstatic, paranormal or religious experience in Great Britain and the United States: a comparison of trends', *Journal for the Scientific Study of Religion* 17:3 (1978), pp. 255–68.
—and Morisy, A., 'Secular society/Religious meanings: a contemporary paradox', *Review of Religious Research*, 26 (1985), pp. 213–27.
—with Nye, R., *The Spirit of the Child*, revised edn (London: Jessica Kingsley Publishers, 2006).
—and Socha, P. M., 'Spirituality as a natural phenomenon: bringing biological and psychological perspectives together', *Zygon* 49:3 (2005), pp. 589–612.
Hayward, J. (ed.), *The Penguin Book of English Verse* (London: Penguin Books, 1956).
Heidegger, M., *Being and Time*, tr. John Macquarrie and Edward Robinson (Oxford: Blackwell, 1962/1927).
Herbert, G., *The Complete English Poems* (London: Penguin Books, 2004).
Himmler, H., 'Speech to the SS Group Leaders on 4th October 1943 in Poznan', taken from the transcript of the Nuremberg Trials, *Trial of the Major War Criminals before the International Military Tribunal*, Nuremberg (1947–9) XXIX, I919-PS.
Hirschman, A., *The Passions and the Interests: Political Arguments for Capitalism Before Its Triumph*, with a foreword by Amartya Sen (Princeton: Princeton University Press, 1997).
Hobbes, T., *Leviathan*, edited and with an introduction by C. B. Macpherson (London: Penguin Classics, 1962).
Horgan, J., *Rational Mysticism: Spirituality Meets Science in the Search for Enlightenment* (Boston and New York: Houghton Mifflin Company, 2003).
Horton, W. M., 'The origin and psychological function of religion, according to Pierre Janet', *American Journal of Psychology* 35:20 (1924).
Huxley, A., *The Perennial Philosophy* (London: Chatto & Windus, 1946)
Huxley, A., *The Doors of Perception* and *Heaven and Hell* (London: Penguin Books, 1954).
Iannaccone, L. R. and Klick, J., 'Spiritual Capital: An Introduction and Literature Review', preliminary draft, prepared for the Spiritual Capital Planning Meeting, 9–10 October 2003, Cambridge, Massachusetts. Available at <http://www.metanexus.net/spiritual_capital/pdf/review.pdf >.
Jackson, M.C., *A Study of the Relationship between Spiritual and Psychotic Experience*, DPhil thesis submitted to Oxford University, 1991.
—and Fulford, K. W. M., 'Spiritual experience and psychopathology', *Philosophy, Psychiatry and Psychology* 4 (1997), pp. 41–90.
James, H. (ed.)., *The Letters of William James* (Boston: Atlantic Monthly Press, 1920).
James, W., *The Principles of Psychology*, 2 vols (New York: Henry Holt in New

York, 1890).
—*The Varieties of Religious Experience* (New York: Longman, Green and Co., 1902).
St John of the Cross, *The Collected Works of St John of the Cross*, tr. Kieran Kavanaugh and Otilio Rodriguez (Washington DC: Institute of Carmelite Studies, 1973).
Johnson, W., *Silent Music: The Science of Meditation* (London: Collins, 1974).
Joravsky, D., *The Lysenko Affair* (Cambridge: Harvard University Press, 1970, republished by the University of Chicago Press, 1986).
Joseph, R., *Neurotheology: Brain, Science, Spirituality, Religious Experience* (San Jose: University Press, 2003).
Kadowaki, J. K., *Zen and the Bible: A Priest's Experience* (London: Routledge and Kegan Paul, 1980).
Kant, E., *Religion Within the Limits of Reason Alone*, tr. Theodore M. Greene and Hoyt H. Hudson (New York: Harper and Row, 1960/1793).
Katz, S. (ed.), *Mysticism and Philosophical Analysis* (New York: Oxford University Press, 1978).
Kenner, H., *A Reader's Guide to Samuel Beckett* (London: Thames and Hudson, 1973).
Kirk, K., Martin, N. and Eaves, L., 'Self-transcendence as a measure of spirituality in a sample of older Australian twins', *Twin Research* 2:2 (1999), pp. 81–7.
Knowlson, J., *Damned to Fame: The Life of Samuel Beckett* (London: Bloomsbury, 1996).
Köhler, W., *The Mentality of Apes*, tr. Ella Winter, 2nd revised edn (London: Routledge and Kegan Paul, 1921).
Kuhn, T., *The Structure of Scientific Revolutions* (Chicago: University of Chicago Press, 1962).
Lakatos, I. and Musgrave, A. (eds), *Criticism and the Growth of Knowledge* (Cambridge: Cambridge University Press, 1970).
Lakoff, G. and Johnson, M., *Metaphors We Live By* (Chicago: University of Chicago Press, 1980).
Lambert, Y., 'A turning point in religious evolution in Europe', *Journal of Contemporary Religion* 19:1 (2004), pp. 29–45.
Larkin, P., *Collected Poems*, edited with an introduction by Anthony Thwaite (London: Faber and Faber, 1988).
Le Bon, G., *The Crowd* (New York: Dover Publications, 2002).
le Mahieu, D. L., *The Mind of William Paley* (Nebraska: University of Nebraska Press, 1976).
Lehodey, V., *The Ways of Mental Prayer*, tr. by a monk of Mount Melleray (Dublin: M. H. Gill and Son Ltd, 1924).
Leroi-Gourhan, A., 'The flowers found with Shanidar IV, a Neanderthal burial in Iraq', *Science* 190 (1975), pp. 562–4.
Leuba, J. H., 'A study in the psychology of religious phenomena', *American Journal of Psychology* 7 (1896), pp. 309–85.
—'Professor William James' interpretation of religious experience', *International Journal of Ethics* 14 (1903–4) pp. 322–9.
—*The Psychology of Religious Mysticism* (London: Kegan Paul, Trench, Trubner and Co., 1925).

—'A psychologist of religion', in Vergilius Ferm (ed.), *Religion in Transition* (London: Allen and Unwin, 1937), pp. 173–200.

Levi, P., *If this is a Man/The Truce*, tr. Stuart Woolf (London: Abacus Books, 1991).

Lindbeck, G., *The Nature of Doctrine: Religion and Theology in a Post-Liberal Age* (Philadelphia: Westminster Press, 1984).

Lobkowicz, N., 'Karl Marx and Max Stirner' in Frederick J. Adelmann (ed.), *Demythologizing Marxism* (Beacon Hill: Boston College, 1969).

Locke, J. L., *Why We Don't Talk to Each Other Any More: the Devoicing of Society* (New York: Simon and Schuster, 1998).

Lukes, S., *Individualism* (Oxford: Blackwell, 1973).

Lumsden, C. and Wilson, E. O., *Promethean Fire: Reflections on the Origin of Mind* (Cambridge, Massachussetts: Harvard University Press, 1983).

Luria, A., *Cognitive Development: Its Cultural and Social Foundations*, tr. Martin Lopez-Morillas and Lynn Solotaroff, ed. Michael Cole (Cambridge, Massachusetts: Harvard University Press, 1976).

Lynch, A., *Thought Contagion: How Ideas Act Like Viruses* (New York: Basic Books, 1998).

Macmurray, J., *The Self as Agent*, with an introduction by Stanley M. Harrison (London: Faber and Faber, 1995).

Macpherson, C. B., *The Political Theory of Possessive Individualism* (Oxford: Oxford University Press, 1962).

Malan, J., 'The possible origin of religion as a conditioned reflex', *American Mercury* 25 (1932), pp. 314–17.

Maratos, O., *The Origin and Development of Imitation in the First Six Months of Life*, PhD thesis submitted to the University of Geneva, 1973.

Maréchal, J., *Studies in the Psychology of the Mystics*, tr. Algar Thorold (London: Burns, Oates and Washbourne, 1927).

Marrett, R., *Head, Heart and Hands in Human Evolution* (London: Hutchinson, 1932).

Marx, K., *Theses on Feuerbach*, reprinted in *Marx and Engels on Religion* (Moscow: Progress Publishers, 1845).

—'The Holy Family' in *Marx and Engels Collected Works*, IV (London: Lawrence and Wishart, 1844).

—and Engels, F., *On Religion* (Moscow: Progress Publishers, 1972).

—*The German Ideology* (Amherst, New York: Prometheus Books, 1998).

Maturin, B., *Laws of the Spiritual Life* (London: Longman, Green and Co., 1909).

Maxwell, M. and Tschudin, V., *Seeing the Invisible* (London: Penguin/Arkana Press, 1990).

McCrone, J., *The Ape that Spoke: Language and the Evolution of the Human Mind* (London: Picador, 1990).

McFague, S., *Metaphorical Theology: Models of God in Religious Language* (London: SCM Press, 1982).

McGrath, A., *Dawkins' God: Genes, Memes and the Meaning of Life* (Oxford: Blackwell, 2005).

Merton, T., *Zen and the Birds of Appetite* (New York: New Directions, 1968).

Miller, P., 'Jonathan Edwards on the Sense of the Heart', *Harvard Theological Review* 41 (1948), p. 124.

Monroe, K. R., *The Heart of Altruism: Perceptions of a Common Humanity* (Princeton: Princeton University Press, 1996).
Morgan Poll, 'Unpublished poll of reports of religious experience in Australia', 1983.
Morris, C., *The Discovery of the Individual, 1050–1200* (London: SPCK, 1972).
Moultrie, G. (1864), 'Let all mortal flesh keep silence' <http://www.cyberhymnal.org/htm/l/e/letallmf.htm>.
Muller, F. Z., *Adam Smith in His Time and Ours: Designing the Decent Society* (New York: The Free Press (Macmillan), 1993).
Muller, F. Max, *Introduction to the Science of Religion* (London: Longmans, Green and Co., 1873).
Murray, L. and Andrews, L., *The Social Baby* (London: CP Publishing, 2000).
Myers, F. W. H., 'The principles of psychology', *Proceedings of the Society for Psychical Research* 7 (1892) p. 301.
Nagy, E. and Molnar, P., 'Homo imitans or Homo provocans? Human imprinting model of neonatal imitation', *Infant Behavior and Development* 27 (2004), pp. 54–63.
Newberg, A., d'Aquili, E. G. and Rause, V., *Why God Won't Go Away: Brain Science and the Biology of Belief* (New York: Ballantine Books, 2001).
Neitz, M. J. and Spickard, J. V., 'Steps toward a sociology of religious experience: the theories of Mihaly Csikszentmihalyi and Alfred Schutz', *Sociological Analysis* 51: 1 (1990), pp. 15–33.
NPDP, *Agreed Minimum Values Framework* (Osborne Park: NPDP, Western Australia, 1995).
Nyanaponika, Thera, *The Heart of Buddhist Meditation* (London: Rider, 1969).
Nye, R. and Hay, D., 'Identifying children's spirituality: how do you start without a starting point?', *British Journal of Religious Education* 18:3 (1996).
Ó'Murchú, D., *Reclaiming Spirituality* (Dublin: Gill and Macmillan, 1997).
Otto, R., *The Idea of the Holy*, 2nd edn, tr. J. W. Harvey (Oxford: Oxford University Press, 1950).
Page, B. (ed.), *Marxism and Spirituality: An International Anthology* (Westport Connecticut: Bergin and Garvey, 1993).
Pahnke, W., *Drugs and Mysticism: An Analysis of the Relationship between Psychedelic Drugs and the Mystical Consciousness.* PhD dissertation, Harvard University, 1963).
Pargament, K., *The Psychology of Religion and Coping: Theory, Research, Practice* (New York, London: Guilford Press, 1997).
Pascal, B., *Pensées*, tr. J. M. Cohen (London: Penguin Classics, 1961).
Paterson, R. W. K., *The Nihilistic Egoist Max Stirner*, published for the University of Hull (Oxford: Oxford University Press, 1971).
Peacocke, A., *Paths from Science towards God* (Oxford: Oneworld Publications, 2001).
Pearse, I. H. and Crocker, L. H., *The Peckham Experiment: a Study in the Living Structure of Society* (London: Allen and Unwin, 1943).
Perls, F. S., *Gestalt Therapy: Excitement and Growth in the Human Personality* (London: Souvenir Press, 1884).
Persinger, M., 'Sense of a presence and suicidal ideation following traumatic brain injury: indications of right hemispheric intrusions from neuropsychological profiles', *Psychological Reports* 75:3 (1994), Pt I, pp. 1059–70.

—'I would kill in God's name: role of sex, weekly church attendance, report of a religious experience and limbic lability', *Perceptual and Motor Skills* 85:1 (1997), pp. 128–130.

—'Religious and mystical experiences as artefacts of temporal lobe function: a general hypothesis', *Perceptual and Motor Skills* 57 (1983), pp. 1235–62.

Popper, K., *The Poverty of Historicism* (London: Routledge and Kegan Paul, 1957).

Porter, R., *Enlightenment: Britain and the Creation of the Modern World* (London: Allen Lane, 2000).

Poulain, A., *The Graces of Interior Prayer* (London: Kegan Paul, Trench, Trubner and Co., 1912).

Preus, S., *Explaining Religion: Criticism and Theory from Bodin to Freud* (New Haven and London: Yale University Press, 1987).

Proudfoot, W., *Religious Experience* (Berkeley: University of California Press, 1985).

—and Shaver, P., 'Attribution theory and the psychology of religion', *Journal for the Scientific Study of Religion* vol. 14:4 (1975), pp. 317–30.

Putnam, R., *Bowling Alone: The Collapse and Revival of American Community* (New York: Simon and Schuster, 2000).

QSR.NUD*IST *User's Guide* (London: Sage/SCOLARI, 1996).

Rahner, K., *Theological Investigations XI*, tr. David Bourke (London: Darton, Longman and Todd, 1974).

Ramachandran, V. S. and Blakeslee, S., *Phantoms in the Brain* (London: Fourth Estate, 1999), ch. 9, 'God and the limbic system', pp. 174–98.

Ramet, S. P. (ed.), *Religious Policy in the Soviet Union* (Cambridge: Cambridge University Press, 1992).

Reid, T., *Thomas Reid's Inquiry and Essays*, ed. Ronald E. Beanblossom and Keith Lehrer (Indianapolis: Hackett Publishing Company Inc., 1983).

Reik, T., *From Thirty Years with Freud* (New York: Farrar and Rinehart, 1941).

Ricœur, P., *The Rule of Metaphor: Multi-disciplinary Studies of the Creation of Meaning in Language*, tr. Robert Czerny (London: Routledge and Kegan Paul, 1978).

Robinson, E., *The Original Vision* (New York: Seabury Press, 1983).

Rogers, C. R., *Client-centred Therapy: Its Current Practice, Implications and Theory* (London: Constable and Robinson, 2003).

Rotter, J. B., *Social Learning and Clinical Psychology* (Englewood Cliffs, New Jersey: Prentice Hall, 1954).

Rottschaefer, W. D., 'The image of God of neurotheology: reflections of culturally based religious commitments or evolutionarily based neuroscientific theories?' *Zygon* 34:1 (1999), pp. 57–65.

Rubenstein, R., *After Auschwitz: History, Theology and Contemporary Judaism*, 2nd edn (Baltimore and London: The Johns Hopkins University Press, 1992).

Saul, R., *The Unconscious Environment* (London: Penguin Books, 1999).

Scherer, K. R., Schorr, A. and Johnstone, T. (eds), *Appraisal Processes in Emotion: theory, methods, research* (Oxford: Oxford University Press, 2001).

Schleiermacher, F., *On Religion: Speeches to its Cultured Despisers*, tr. John Oman (New York: Harper and Row, 1958).

Schmid, H., *The Doctrinal Theology of the Evangelical Lutheran Church*, tr.

Charles A. Hay and Henry E. Jacobs (Minneapolis: Augsburg Publishing House, 1961/1876).

Schutz, A., 'Making music together: a study in social relationship', in Arvid Brodersen (ed.), *Collected Papers II: Studies in Social Theory* (The Hague: Martinus Nijhoff, 1964).

Selznick, P., *The Moral Commonwealth: Social Theory and the Promise of Community* (Berkeley: University of California Press, 1992).

Sensky, T., Wilson, A., Petty, R., Fenwick, P. B. C. and Rose, F. C., 'The interictal personality traits of temporal lobe epileptics: religious belief and its association with reported mystical experiences', in Porter, R. J. *et al.* (eds), *Advances in Epileptology: XVth Epilepsy International Symposium* (New York: Raven Press, 1984), pp. 545–9.

Shattuck, R., *Forbidden Knowledge: from Prometheus to Pornography* (San Diego/New York/London: Harvest Books, 1997).

Shreeve, J., *The Neanderthal Enigma: Solving the Mystery of Modern Human Origins* (New York: William Morrow and Company Inc., 1995).

Sidis, B., *The Psychology of Suggestion* (New York: D. Appleton, 1903).

Smith, A., *The Wealth of Nations*, 2 vols., with an introduction by Andrew Skinner (London: Penguin Books, 1999).

—*The Theory of Moral Sentiments* (Amherst, New York: Prometheus Books, 2000/1759).

Smith, H., 'Do drugs have religious import?', *Journal of Philosophy* 61:18 (1964), pp. 517–30.

Smuts, J. C., *Holism and Evolution* (London: Macmillan, 1926).

Solecki, R., *Shanidar: The First Flower People* (New York: Knopf, 1971).

Soskice, J. M., *Metaphor and Religious Language* (Oxford: Clarendon Press, 1987).

Stace, W., *Mysticism and Philosophy* (London: Macmillan, 1960).

Starbuck, E. D., *The Psychology of Religion* (London: Walter Scott, 1899).

—'Religion's use of me', in Vergilius Ferm (ed.), *Religion in Transition* (London: Allen and Unwin, 1937), p. 226.

Steiner, G., *The Death of Tragedy* (London: Faber and Faber, 1962).

—'"Tragedy" reconsidered', *New Literary History* 35 (2004), pp. 1–15.

Stiglitz, J., *Globalization and its Discontents* (London: Penguin Books, 2002).

Stirner, M., *The Ego and Its Own*, tr. Steven Byington, with an introduction by Sydney Parker (London: Rebel Press, 1993).

Stolz, A., *The Doctrine of Spiritual Perfection* (St Louis: B. Herder Book Co., 1938).

Stratton, G. M., *Psychology of the Religious Life* (London: Allen and Unwin, 1911).

Stringer, C. B., 'Evolution of early humans' in *The Cambridge Encyclopedia of Human Evolution*, Steve Jones, Robert Martin and David Pilbeam (eds) (Cambridge: Cambridge University Press 1994), pp. 241–51.

Strout, C., 'The pluralistic identity of William James', *American Quarterly* 23:2 (1971), p. 135

Tacey, D., *The Spirituality Revolution: the emergence of contemporary spirituality* (Hove and New York: Brunner-Routledge, 2004).

Tamminen, K., *Religious Development in Childhood and Youth: An Empirical Study*, (Helsinki: Suomalainen Tiedeakatemia, 1991).

—'Religious experiences in childhood and adolescence: a viewpoint of religious development between the ages of 7 and 20', *International Journal for the Psychology of Religion* 4:2 (1994), pp. 61–85.

Tanquerey, A., *The Spiritual Life: A Treatise on Ascetical and Mystical Theology*, tr. Herman Branderis (Tournai/Paris/Rome/New York: Desclee and Co., 1930).

Taylor, C., *Sources of the Self: The Making of the Modern Identity* (Cambridge: Cambridge University Press, 1989).

Taylor, C., *Varieties of Religion Today: William James Revisited* (Cambridge, Massachusetts: Harvard University Press, 2002).

Thiselton, A., *New Horizons in Hermeneutics* (London: HarperCollins, 1992).

Trevarthen, C., 'The self born in intersubjectivity, an infant communicating', in U. Neisser (ed.), *The Perceived Self: Ecological and Interpersonal Sources of Self Knowledge* (New York: Cambridge University Press, 1993).

—'Proof of sympathy: Scientific evidence on the co-operative personality of the infant, and evaluation of John Macmurray's "Mother and Child"', in David Fergusson and Nigel Dower (eds), *John Macmurray: Critical Perspectives*, Proceedings of the Conference at Kings College, Old Aberdeen, 6–9 April, 1998, (New York: Peter Lang, 1998).

—'The neurobiology of early communication: intersubjective regulation in human brain development', in A. F. Kalverboer and A. Grimsbergen (eds), *Handbook of Brain and Behavior in Human Development* (Dordrecht: Kluwer Academic Publications, 2001).

Troeltsch, E., *The Social Teaching of the Christian Churches*, 3 vols, tr. Olive Wyon (London: Allen and Unwin, 1931).

Trotter, J. D., *Wholeness and Holiness: A Study in Human Ethology and the Holy Trinity* (Glasgow: Pioneer Health Foundation, 2003).

United States Holocaust Memorial Museum's website: <http://www.ushmm.org/wlc/article.php?lang=enand ModuleId=10005537>.

Van Ness, P. H., *Spirituality and the Secular Quest* (London: SCM Press Ltd, 1996).

Voltaire, *Candide*, ed. Haydn Mason (London: Bristol Classical Press, 1995).

von Hartmann, E., *Das Religiöse Bewussstsein der Menschheit im Stufengang seiner Entwicklung* (Berlin: Carl Buncker, 1882).

Vygotsky, L., *Thought and Language*, 2nd edn (Cambridge, Massachusetts: MIT Press, 1882).

Wach, J., *The Comparative Study of Religions* (New York: Columbia University Press, 1958).

Walzer, M., 'The communitarian critique of liberalism', *Political Theory* 18:1 (1990), pp. 6–23.

Ward, M., *Father Maturin: A Memoir* (London: Longman, Green and Co., 1920).

Watson, J., *The Double Helix* (London: Penguin Books, 1968).

Weber, M., *The Protestant Ethic and the Spirit of Capitalism*, tr. Talcott Parsons (London: Allen and Unwin, 1930).

Wheen, F., *Karl Marx* (London: Fourth Estate, 2000).

White, L. A., 'Individuality and Individualism: a culturological interpretation', *Texas Quarterly* 6:2 (1963), pp. 111–27.

Wilde, L., *Ethical Marxism and its Radical Critics* (London: Macmillan, 1998).

Wilson, B., *Religion in Secular Society* (London: C. A. Watts, 1966).
Wordsworth, W., *The Prelude* (Oxford: Oxford University Press, 1977).
Wulff, D., *Psychology of Religion: Classic and contemporary*, 2nd edn (New York: Wiley and Sons, 1997).
Wuthnow, R., *The Consciousness Reformation* (Berkeley: University of California Press, 1976).
Yeats, W. B., *The Collected Poems of W. B. Yeats* (Ware: Wordsworth Editions, 2000).
Yoshikawa, M. J., 'The double swing model of intercultural communication between the East and the West', in D. Lawrence Kincaid (ed.), *Communication Theory: Eastern and Western Perspectives* (San Diego: Academic Press, 1987).
Zimmer, C., 'Faith boosting genes: a search for the genetic basis of spirituality', *Scientific American*, 27 September 2004.
Zinnbauer, B. J., Pargament, K., Cole, B., Rye, M., Butter, E., Belavich, T., Hipp, K., Scott, A. and Kadar, J., 'Religion and spirituality: unfuzzying the fuzzy', *Journal for the Scientific Study of Religion*, 76:4 (1997), pp. 549-64.
Zohar, D. and Marshall, I., *Spiritual Capital: Wealth We Can Live By* (London: Bloomsbury, 2004).

Index